Lecture Notes in Physics

T0242454

The Lecture Notes in Physics

The series Lecture Notes in Physics (LNP), founded in 1969, reports new developments in physics research and teaching – quickly and informally, but with a high quality and the explicit aim to summarize and communicate current knowledge in an accessible way. Books published in this series are conceived as bridging material between advanced graduate textbooks and the forefront of research and to serve three purposes:

- to be a compact and modern up-to-date source of reference on a well-defined topic

- to serve as an accessible introduction to the field to postgraduate students and nonspecialist researchers from related areas

- to be a source of advanced teaching material for specialized seminars, courses and schools

Both monographs and multi-author volumes will be considered for publication. Edited volumes should, however, consist of a very limited number of contributions only. Proceedings will not be considered for LNP.

Volumes published in LNP are disseminated both in print and in electronic formats, the electronic archive being available at springerlink.com. The series content is indexed, abstracted and referenced by many abstracting and information services, bibliographic networks, subscription agencies, library networks, and consortia.

Proposals should be sent to a member of the Editorial Board, or directly to the managing editor at Springer:

Christian Caron
Springer Heidelberg
Physics Editorial Department I
Tiergartenstrasse 17
69121 Heidelberg / Germany
christian.caron@springer.com

J. Gracia
F. De Colle
T. Downes (Eds.)

Jets from Young Stars V

High Performance Computing and Applications

 Springer

MARIE CURIE ACTIONS

José Gracia
MPI für Astronomie
Königstuhl 17
69117 Heidelberg
Germany
gracia@hlrs.de

Fabio De Colle
Dublin Institute for Advanced Studies
School of Cosmic Physics
5 Merrion Square
Dublin 2
Ireland
fabio@ucolick.org

Turlough Downes
Dublin City University
School of Mathematical Sciences
Glasnevin
Dublin 9
Ireland
turlough.downes@dcu.ie

Gracia J. et al. (Eds.), *Jets from Young Stars V: High Performance Computing and Applications*, Lect. Notes Phys. 791 (Springer, Berlin Heidelberg 2009),
DOI 10.1007/978-3-642-03370-4

Lecture Notes in Physics ISSN 0075-8450   e-ISSN 1616-6361
ISBN 978-3-642-26130-5   e-ISBN 978-3-642-03370-4
DOI 10.1007/978-3-642-03370-4
Springer Heidelberg Dordrecht London New York

Cover design: Integra Software Services Pvt. Ltd., Pondicherry

Printed on acid-free paper

Springer is part of Springer Science+Business Media (www.springer.com)

Preface

One of the most interesting and spectacular phenomena in star-forming regions is the presence of extended and well-collimated jets of material moving away from the protostar. The study of the formation of these jets and their propagation provides important clues about the early stages of stellar evolution. In particular, it is possible to study their key role in the extraction of angular momentum from the collapsing gas cloud and their effect on the internal structure of protoplanetary disks. In order to understand the dynamics of the accretion disk–star system and the processes responsible for the ejection, collimation, and evolution of outflows from these systems, it is necessary to solve a set of differential equations. Due to the complexity of this system of equations, there is only a small number of analytical solutions available. Often their very nature limits their application to real astrophysical systems. Therefore numerical methods are necessary to understand the majority of the physical processes involved.

Modern observational facilities, on the other hand, are now producing huge amounts of data often requiring considerable computational resources to fully exploit them. In most cases this data becomes publicly available within a reasonably short time. Virtual observatories allow access not only to public data archives, but also allow to interface coherently different repositories running on distributed hardware, i.e., Computational Grids.

JETSET (JET Simulations Experiments and Theory) is a four-year Marie Curie Research Training Network (RTN) designed to build a vibrant interdisciplinary European Research and Training community centered on rigorous and novel approaches to plasma jet studies, with a focus on flows produced during star formation. The theme of the network is at the confluence of astrophysical observations, theoretical and computational modeling, laboratory experiments, and Grid technology. The network scientific goals focused on understanding (i) the driving mechanisms of jets around young stars including their possible link with planet-forming disks; (ii) the cooling–heating processes, instabilities, and shock structures in stellar and laboratory jets; (iii) and the impact of jets on energy balance and star formation in the galactic medium. Central to these overall goals are the series of JETSET schools dedicated to the training of our young researchers in key jet topics.

The first school (Villard-de-Lans, France, January 2006) was focused on the magnetohydrodynamic (MHD) theoretical models used to describe jets from young stars, while the second school (Marciana Marina, Elba Island, Italy, September 2006) was devoted to the current techniques to observe Jets, Outflows, and Circumstellar Disks at high angular resolution, and on the procedures applicable to the data towards an initial interpretation. The third school (Sauze d'Oulx, Italy, January 2007) had as its main topic "Numerical MHD and Instabilities", with a special session dedicated to Visualization techniques and virtual reality. Finally, the fourth school (Azores, Portugal, June 2007) aimed at bridging the gap between models, observations, and experiments of jets.

This book is a collection of the lectures from the fifth and last school of the JETSET network, "Jets From Young Stars: High Performance Computing in Astrophysics", held in Galway, Ireland, in January 2008. Stellar jets are complex physical systems. Their study necessitates incorporating nonlinear effects which occur on a wide variety of length and timescales. As a result of this one of the primary methods used to study the physics of jets is numerical simulations. Since the physics is so complex, the problems are large enough to make mandatory the use of high-performance computing techniques. Further, modern astrophysical datasets associated with observations are becoming so large that standard analysis methods are no longer efficient. The aim of this school was, therefore, to address the methods used to simulate astrophysical jets and to derive useful information from large datasets, focusing on techniques involving high-performance computing.

The first part of the book is devoted to general aspects of high performance in astrophysics. First, Honoré Tapamo introduces us to parallel techniques, with emphasis on MPI. John Walsh reviews grid technology techniques. The chapter by Nicholas Walton is a practical introduction to Virtual Observatory (with particular reference to AstroGrid). The second part of the book is devoted to applications of high-performance computing techniques to jet and star formation processes. Roby Banerjee describes the processes leading to the formation of jets from the collapse of magnetized cloud cores. Jürgen Steinacker shows the techniques underlying a three-dimensional radiation transfer code, and its application to astrophysical problems. Claudio Zanni, Rony Keppens, and Turlough Downes illustrate the three fundamental aspects of the stellar jet phenomena: jet ejection and its relation to the star/disk interaction (Claudio Zanni), jet stability (Rony Keppens) and the large scale propagation of jets and consequent interaction with the ambient medium (Turlough Downes).

The editors would like to thank all the lecturers for their excellent presentations and contributions to this book. We are also thankful to all school participants who, in collaboration with the lecturers, made the school an enjoyable, exciting, and informative occasion for everyone involved.

We would like to aknowledge the other members of the scientific committee: Sylvie Cabrit, Max Camenzind, Catherine Dougados, and Tom Ray for their help in organizing the scientific aspects of the school and Eileen Flood, Matt Redman, and

Emma Whelan for their superb contribution to the organization of all the practical aspects of this school.

Heidelberg, Germany   José Gracia
Dublin, Ireland   Fabio De Colle
Dublin, Ireland   Turlough Downes
                                                                         December, 2008

Contents

Part II Applications in Astrophysics

Part I
High Performance Computing

Introduction to Message-Passing Interface

Honoré Tapamo

Abstract This introduction to the Message-Passing Interface (MPI) course provides an overview of the MPI standard with a view to its application to scientific/engineering problems. Throughout the course, the basic concepts behind the message-passing paradigm are discussed together with code examples. Readers will also find practical excercises at the end of each section, where they will have the opportunity to explore motivational examples in C or Fortran.

1 Introduction

The course is subdivided into eight sections each covering a particular feature of MPI-1. At the end of each section a practical exercise is given emphasizing the main MPI concept studied in that section. The organization is the following.

Section 2 gives an overview of MPI. Here the definition and the goal of MPI is given. The Message-passing programming paradigm is introduced. The concept of Data and Work Distribution and MPI-targeted platforms is introduced. Readers will learn that MPI is not a new language but rather a collection of functions with C/C++ and Fortran bindings. They will learn how to compile MPI programs using wrappers. At the end of this section readers will know what MPI is all about and will also be able to know if they can benefit from MPI.

Section 3 discusses the process model and language binding. Here readers will learn how to call MPI routines within a program. The format of MPI routines is presented. The concept of handle is briefly introduced to present the default communicator. Basics routines needed to write a minimal MPI program are introduced. In the practical excercise at the end of this section, readers will write a simple program to print *hello world*. They will learn how to initialize and exit MPI communications. They will also learn how to query the processors identifiers and the number of processes.

H. Tapamo (✉)
Irish Center for High End Computing, Dublin, Ireland, htapamo@ichec.ie

Tapamo, H.: *Introduction to Message-Passing Interface*. Lect. Notes Phys. **791**, 3–47 (2009)
DOI 10.1007/978-3-642-03370-4_1   © Springer-Verlag Berlin Heidelberg 2009

Section 4 introduces the concept of messages and point-to-point communication. This section starts with the description of a message. MPI datatypes are also introduced and the list of basic MPI datatypes is presented. Then the simplest form of communication, namely the point-to-point communication, is introduced. Two MPI routines, *MPI_Send* and *MPI_Recv*, needed for point-to-point communications are studied in detail. The concept of communication mode for communication optimization is also introduced. The practical exercise at the end of this section is about the use of *MPI_Send* and *MPI_Recv* routines to send and receive data between two processors. The advanced exercise introduces the concepts of bandwidth and latency.

Section 5 introduces the concept of nonblocking communications. First of all the mechanism of separating the communication in two phases is explained. Each phase is explained in detail. The concept of hiding the latency and avoiding deadlock is shown in various examples. MPI communication routines associated to each phase are presented. The concept of request handle is mentioned here to introduce the mechanism of message completion. The practical exercise in this section deals with rotating information around a ring. The processors are logically arranged in a ring. Each processor stores its rank in a variable and pass it to its right neighbor which forward this value further. Nonblocking send and receive should be used here to avoid a deadlock and to verify correctness because blocking synchronous send will cause a deadlock.

Section 6 deals with collective communications. The straightforward philosophy behind collective communication is explained; then emphasis is made on the fact that collective communication routines are optimized and should be used whenever and wherever possible. The most commonly used collective routines are presented and discussed. The practical exercise here consists in rewriting the program in Section 4 but using the MPI global reduction to perform the global sum of all ranks of the processes in the ring.

Section 7 is about communicators, groups, and virtual topologies. In the previous sections the concept of communicator has been vaguely introduced; here more details are given. It is shown that one can create groups of processors. Routines to associate communicators to these groups are presented. Now having new communicators one can use them as one used to use the default communicator. MPI routines needed to map the abstract topology on to the natural topology of the problem domain are presented. The rest of this section presents the MPI routines used to manipulate two-dimensional Cartesian topology.

Section 8 introduces the concept of derived datatypes. So far one has used basic datatypes. It is shown here how MPI deals with complex datatypes. MPI-derived datatypes associated to commonly used complex datatypes are presented. The concept of data layout is introduced to ease the understanding of MPI-derived datatypes philosophy. Some examples are shown for the *MPI_TYPE_Vector* routine since it is one of the mostly used.

Section 9 is a case study about the parallel implementation of the advection equation. The aim of the case study is to apply all the MPI concepts seen during the course on a real world problem. First students have to use what they learned in Section 6 to create a two-dimensional Cartesian topology, then they will partition the two-dimensional array between processors. They will create a master–slave model and map the virtual Cartesian topology on to the two-dimensional array partition (using again concepts introduced in Section 6). They will create MPI-derived datatypes (introduced in Section 7) to swap (send and receive) halo regions between neighboring processors.

2 MPI Overview

MPI defines a standard for message-passing interface and is the result of the effort of many people around the world. Message passing is a paradigm used widely on certain classes of parallel machines, especially those with distributed memory. Although there are many variations, the basic concept of processes communicating through messages is well understood. Over the last ten years substantial progress has been made in casting significant applications in this paradigm. Each vendor has implemented its own variant. Among these distributions we can cite MPICH, LAM and OpenMPI. MPI specifies a standard for parallel computers, clusters and heterogeneous networks. It is not a specific product nor a new programming language. It is simply a series of libraries of many functions, portable with Fortran and C/C++ interfaces. Debugging MPI programs can be very difficult due the MPMD (Multiple Programs Multiple Data) emulation that they usually imply.

2.1 Goals and Scope of MPI

MPI's prime goals are to

- Provide a message-passing interface.
- Provide source-code portability.
- Allow efficient implementations

It also offers:

- A great deal of functionality.
- A support for heterogeneous parallel architectures.

With MPI-2 one has

- Important additional functionality.
- Backwards compatible with MPI-1.

2.1.1 Compilation and Parallel Start

Using wrappers provided by MPI, one can compile its program as usual. The advantage with these wrappers is that the programmer does not have to explicitly provide the path for the header files and the libraries. Given your programming language choice and the architecture you have the following options for the machines assigned for the present course.

– Compilation in C: **mpicc** -o prog prog.c
– Compilation in C++: **mpiCC** -o prog prog.cpp
– Compilation in Fortran **mpif77, mpif90** -o prog prog.f90

Other wrappers might be available depending on the specific machine/architecture, like **mpixlc, mpixlf90** on the blue gene architectures.

One calls a special command to start parallel programs. Again, according to your architectures you will use **mpirun, mprun** or **mpiexec**. For instance, the command

```
mprun -n num ./prog
```

starts the parallel program **prog** on **num** processors.

2.1.2 Message-Passing Programming Paradigm

In the sequential programming paradigm one has one program running on data in the same memory space, see Fig. 1.

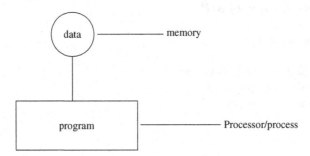

Fig. 1 Sequential programming paradigm, single memory space but a processor may run many process

In the message-passing paradigm the sequential programming paradigm is replicated many times. In the message-passing one usually has many processors running the same program on a different data (memory space). Processors can exchange data via a communication network, see Fig. 2.

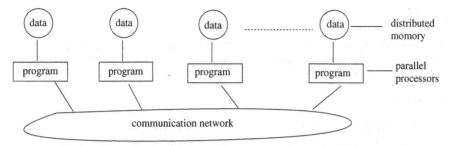

Fig. 2 Message-passing programming paradigm, multiple memory spaces

A **process** is a program performing a task on a processor; each processor/process in a message-passing program runs a instance/copy of a program. Hence, each processor runs a sub-program characteristics of which are listed below.

- The sub-program is written in a conventional sequential language, e.g., C or Fortran.
- The sub-program is typically a single program operating on multiple dataset.
- Variables of each sub-program have

  - the same name,
  - but different locations (distributed memory) and different data, i.e., all variables are local to a process.

- Sub-programs communicate via a special send and receive routines (**message passing**).

2.1.3 Data and Work Distribution

To communicate together MPI-processes need identifiers usually called **ranks**. The rank is the identifying number on which all distribution decisions are based.

MPI allows the emulation of **MPMD**, i.e., Multiple Program Multiple Data. As opposed to **SPMD** (Single Program Multiple Data) where the same sub-program runs on each processors, MPMD allows the execution of different programs on different data. The emulation of MPMD is done in MPI via branching in the program, see below.

```
main(int argc, char **argv){
   if (myrank < .... /* process should run
        the ocean model */){
     ocean( /* arguments */ );
   }
   else{
     weather( /* arguments */ );
   }
}
```

2.2 Message Passing

To communicate, MPI processes send and receive messages. Messages are packets of data moving between sub-programs. The necessary information for a successful message passage are

- From the sender:

  - sending process rank (identifier of the sending process),
  - source location (the address of the variables containing the data to send),
  - source datatype (the datatype of the variable containing the data to send),
  - source data size (the number of elements to send).

- From the receiver:

  - receiving process (identifier of the receiving process),
  - destination location (the address of the variable where the data will be stored),
  - destination data size (the number of elements to receive).

These information constitute the "envelop" of a message, a concept that will be clarified in Section 4.

2.3 MPI Communication Forms

There are two forms of communication in MPI-1, namely point-to-point communication and collective communication.

2.3.1 Point-to-Point Communication

The point-to-point communication is the simplest form of message passing involving two processes at the same time only. One process sends a message to another. One distinguishes two types of message sending: the synchronous and the buffered send. In the synchronous send, the sender gets confirmation that the message is received. It is analogue to the beep or okay-sheet of a fax. In the buffered or asynchronous send the sender knows only when the message has left; it does not wait for any information about the reception of the message.

There are also two types of communications within MPI. The **blocking** and **non-blocking operations**. The synchronous send operation is blocking; the sender waits (blocks) until a receive operation is issued. Blocking subroutines return only when the operation has completed. We will clarify these sometimes confusing but important MPI concepts in Section 5.

2.3.2 Collective Communications

As opposed to point-to-point communications, collective communications involve many processors at the same time. They are higher level communication routines that may allow optimized internal implementation, e.g., tree-based algorithms. Among the mostly used collective operations, one can cite:

- The **broadcast** operation (one-to-many communication). Here one processor sends the same message to other processes.
- The **reduction** operation which consist of combining data from several processes to produce a single result.
- The **barrier** which synchronizes the processes at a particular point in the program.

3 Process Model and Language Bindings

In this section we will present basic MPI routines and show their language bindings. MPI routines are declared in mpi.h for C/C++ programs and mpif.h for Fortran programs. The general function format is

- *error = MPI_Xxxxx(parameter,...)* in C.
- *CALL MPI_XXXXX(parameter,* **IERROR***)* in Fortran.

All MPI routines are defined in the standard. They are language independent. *MPI_* namespace is reserved for MPI constants and routines, i.e., application routines and variable names must not begin with *MPI_*.

3.1 Initializing MPI

MPI_Init is the first routine that must be called in an MPI program. Every MPI program must call this routine once, before any other MPI routines. Making multiple calls to *MPI_INIT* is erroneous. The C version of the routine accepts the arguments to main, argc and argv as arguments.

```
int MPI_Init(int *argc, char ***argv);
```

The Fortran version takes no arguments other than the error code.

```
MPI_INIT(IERROR), INTEGER IERROR
```

3.2 The MPI_COMM_WORLD Communicator

All MPI routines need a communicator. In other words every communication is relative to a communicator. All processors are members of a communicator called *MPI_COMM_WORLD*. A communicator is simply a mechanism used within MPI to identify a group of processors; more details about that will be given in Section 3.

MPI_COMM_WORLD is a predefined object in mpi.h and mpif.h. Each processor is identified by a rank in a communicator, starting with 0 and ending with size − 1, see Fig. 3.

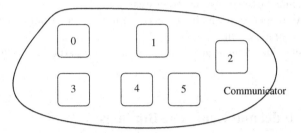

Fig. 3 The *MPI_COMM_WORLD* communicator in a program using six processors

3.3 Clean-up of MPI

An MPI program should call the MPI routine *MPI_FINALIZE* when all communications have completed. This routine cleans up all MPI data structures, etc. It does not cancel outstanding communications, so it is the responsibility of the programmer to make sure all communications have completed. Once this routine has been called, no other calls can be made to MPI routines, not even *MPI_INIT*, so a process cannot later re-enrol in MPI.

3.3.1 First Example

A simple MPI program would look like this:

C version

```
#include <mpi.h>
/* Also include usual header files */
int main(int argc, char **argv)
    {
        /* Initialise MPI */
        MPI_Init (&argc, &argv);
        /* There is no main program */
        /* Terminate MPI */
        MPI_Finalize ();
        return 0;
    }
```

Fortran version

```
PROGRAM simple
   include mpif.h
   integer errcode
C Initialise MPI
   call MPI_INIT (errcode)
C The main part of the program goes here.
C Terminate MPI
   call MPI_FINALIZE (errcode)
   end
```

3.3.2 Accessing Basic Communicator Information

One of the basic information a programmer would like to know is the rank of each processor and the number of processors involved in the parallel process. This is achievable using the *MPI_COMM_RANK* and *MPI_COMM_SIZE* routines.

```
MPI_COMM_RANK(comm,rank)
```

returns in *rank* the rank of the calling process associated to the communicator *comm*. Remember: every communication is relative to a communicator!

```
MPI_COMM_SIZE(comm,rank)
```

returns in *size* the number of processor associated to the communicator comm.

3.3.3 Exercise: Hello World

1. Write a minimal MPI program which prints the message "Hello World!". Compile and run it on a single processor.
2. Run it on several processors in parallel.
3. Modify your program so that only the process ranked 0 in *MPI_COMM_WORLD* prints out the message.
4. Modify your program so that

   - every process writes its rank and the size of *MPI_COMM_WORLD*,
   - only process ranked 0 in *MPI_COMM_WORLD* prints "hello world!".

5. Why is the sequence of the output non-deterministic?
6. What must be done to ensure that the output of all MPI processes is printed to the terminal window in rank-sequence? Or is it not guaranteed?

4 Messages and Point-to-Point Communication

This is the simplest form of communication in the sense that it involves two processes at a given time. One process sends and the other process receives a **message**.

4.1 What is a Message?

An MPI message contains a number of elements of some particular datatype. All
MPI messages are typed in the sense that the type of the contents must be specified
in the send and receive. See Fig. 4.

2345	654	96574	−12	7676

Fig. 4 Example: message with five integers

Derived datatypes are built from basic datatypes which correspond to the basic C
and Fortran datatypes as shown in the Tables 1 and 2 below.

Derived datatypes can be constructed at run-time; they can be used for sending
strided vectors, C structs, etc.

Table 1 Basic C datatypes in MPI

MPI datatype	C datatype
MPI_CHAR	Signed char
MPI_SHORT	Signed short int
MPI_INT	Signed int
MPI_LONG	Signed long int
MPI_UNSIGNED_CHAR	Unsigned char
MPI_UNSIGNED_SHORT	Unsigned short int
MPI_UNSIGNED	Unsigned int
MPI_UNSIGNED_LONG	Unsigned long int
MPI_FLOAT	Float
MPI_DOUBLE	Double
MPI_LONG_DOUBLE	Long double
MPI_BYTE	
MPI_PACKED	

Table 2 Basic fortran datatypes in MPI

MPI datatype	Fortran datatype
MPI_INTEGER	INTEGER
MPI_REAL	REAL
MPI_DOUBLE_PRECISION	DOUBLE PRECISION
MPI_COMPLEX	COMPLEX
MPI_LOGICAL	LOGICAL
MPI_CHARACTER	CHARACTER(1)
MPI_BYTE	
MPI_PACKED	

There are rules for datatype-matching and, with certain exceptions, the datatype specified in the receive must match the datatype specified in the send. The great advantage of this is that MPI can support heterogeneous parallel architectures, i.e., parallel machines built from different processors, because type conversion can be performed when necessary. Thus two processors may represent, say, an integer in different ways, but MPI processes on these processors can use MPI to send integer messages without being aware of the heterogeneity.

4.2 Sending a Message

A **point-to-point communication** as opposed to **collective** communication involves exactly two processes. One process sends and the other receives. See Fig. 5.

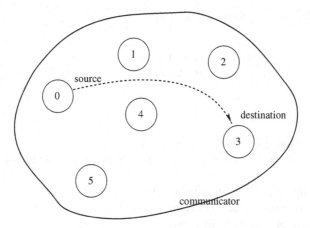

Fig. 5 Example of point-to-point communication, process with rank 0 (the source) sends a message to processor with rank 3 (the destination)

The standard send has the following form.

```
MPI_SEND (buf, count, datatype, dest, tag, comm),
```

where

- **buf** is the starting point of the message with **count** elements, each described with **datatype**.
- **dest** is the rank of the destination process within the communicator **comm**.
- **tag** is an additional nonnegative integer piggyback information, additionally transferred with the message. It can be used by the program to distinguish messages.

Below are the C and Fortran interface of the standard *MPI_Send* routine. Note that the Fortran routine takes a last output argument which gives the status of the sending operation. Such extra argument is present in all MPI routines, expect *MPI_WTIME* and *MPI_WTICK*.

```
C: int MPI_Send(void *buf, int count, MPI_Datatype
               datatype, int dest, int tag,
               MPI_Comm comm)

Fortran:MPI_SEND(BUF, COUNT, DATATYPE, DEST,
                TAG, COMM, IERROR)
<type> BUF(*)
INTEGER COUNT, DATATYPE, DEST,
               TAG, COMM, IERROR
```

4.3 Receiving a Message

The standard receive has the following form

```
MPI_RECV(buf, count, datatype, source,
         tag, comm, status)
```

where

- **buf/count/datatype** describe the receive buffer.
- This routine is used to receive a message sent by process with rank **source** in **comm**.
- **Envelope** information is returned in **status**.
- Only messages with matching **tag** are received.
- buf and status are output arguments.

4.4 Requirements for Point-to-Point Communications

For a communication to succeed

- Sender must specify a valid destination rank. In Fig. 5, a valid rank is an integer from 0 to 5.
- Receiver must specify a valid source rank. In Fig. 5, a valid rank is an integer from 0 to 5.
- The communicator must be the same.
- Tags must match. If you specify different tags then the receive operation will not complete.
- Message datatypes must match. One has to use the same datatype for sending and for receiving.
- Receiver's buffer must be large enough. This means that the buffer used to receive a message should be larger than the number of bytes sent.

4.4.1 Wildcards

The receiver in a MPI receiving routine can use wildcards. That is a process can receive a message from any source or with any tag in which case the **source** in the *MPI_Recv* should be set to *MPI_ANY_SOURCE* and the **tag** should be set to *MPI_ANY_TAG*. In this case the actual source and tag are returned in the receiver's **status** parameter and this constitutes the envelope of the message. The status output parameter is a structure in C/C++ and an array in Fortran whose elements are actual source and tag of the message. In C/C++ for example status. *MPI_SOURCE* and status. *MPI_TAG* would return the source and the tag of the message received using the status parameter. In Fortran the same information is obtained from the variables status(*MPI_SOURCE*) and status(*MPI_TAG*).

MPI also provides the function *MPI_get_count( )* which returns the number of elements received.

4.5 Communication Modes

The send call described earlier used the standard communication mode. MPI provides the programmer with three other communication modes: **synchronous**, **buffered** and **ready** modes. These modes correspond to different types of send. The communication mode is meaningful only for the sender and not for the receiver. In short, the *completion* of a send means that the send buffer can safely be reused. The difference between these communication modes resides in how the completion of a send depends on the reception of the message. Table 3 below gives the definition of the different communication modes.

Table 3 MPI communication modes explained

Sender mode	Definition (completion condition)
Synchronous send **MPI_SSEND**	Completes when the receive has started
Buffered send **MPI_BSEND**	Always completes (unless an error occurs), irrespective of receiver
Ready send **MPI_RSEND**	May be started only if the matching receive is already posted
Receive **MPI_Recv**	Completes when the message (data) has arrived

A **buffered** mode send operation can be started whether or not a matching receive has been posted. It may complete before a matching receive is posted. However, unlike the standard send, this operation is local, and its completion does not depend on the occurrence of a matching receive. Thus, if a send is executed and no matching receive is posted, then MPI must buffer the outgoing message, so as to allow the send call to complete. An error will occur if there is insufficient buffer space. The amount of available buffer space can be controlled by the user. Buffer allocation by the user may be required for the buffered mode to be effective.

A send that uses the **synchronous** mode can be started whether or not a matching receive was posted. However, the send will complete successfully only if a matching receive is posted, and the receive operation has started to receive the message sent by the synchronous send. Thus, the completion of a synchronous send not only indicates that the send buffer can be reused, but also indicates that the receiver has reached a certain point in its execution, namely that it has started executing the matching receive. If both sends and receives are blocking operations then the use of the synchronous mode provides synchronous communication semantics: a communication does not complete at either end before both processes rendezvous at the communication.

A send that uses the **ready** communication mode may be started only if the matching receive is already posted. Otherwise, the operation is erroneous and its outcome is undefined. On some systems, this allows the removal of a hand-shake operation that is otherwise required and results in improved performance. The completion of the send operation does not depend on the status of a matching receive, and merely indicates that the send buffer can be reused. A send operation that uses the ready mode has the same semantics as a standard send operation, or a synchronous send operation; it is merely that the sender provides additional information to the system (namely that a matching receive is already posted), that can save some overhead. In a correct program, therefore, a ready send could be replaced by a standard send with no effect on the behavior of the program other than performance.

4.5.1 Message Order Preservation

The order of messages in MPI is preserved. It means that if two messages have the same envelope and are using the same communicator, then a message sent first will be received first: messages do not overtake each other; see Fig. 6.

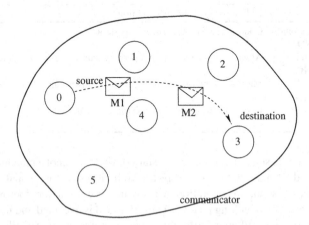

Fig. 6 Example of two messages with the same envelope: M_1 will be received before M_2; there is no overtake

4.5.2 Exercise: Ping-Pong

- Write a program according to the time-line diagram in Fig. 7 below.

  – process 0 sends a message to process 1 (ping)
  – after receiving this message, process 1 sends a message back to process 0 (pong)

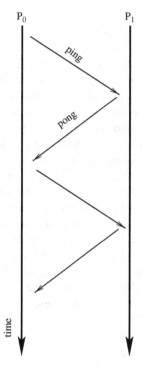

Fig. 7 Illustration of a the ping pong exercise: P_0 sends to P_1 who receives and sends back

- Repeat this ping-pong with a loop of length 50
- Add timing calls before and after the loop using the function *MPI_Wtime()*
- At process 0, print out the transfer time of one message

  – in second
  – in microsecond

- Print out message transfer time and bandwidth

  – for the following send modes:

    - for standard send (*MPI_Send*),
    - for synchronous send (*MPI_Ssend*),

  – and for the following message sizes:

- 8 bytes (e.g., one double or double precision value),
- 512 B (= 8*64 bytes),
- 32 kB (= 8*64**2 bytes),
- 2 MB (= 8*64**3 bytes).

Latency = transfer time for zero length messages.
bandwidth = message size (in bytes) / transfer time

5 NonBlocking Communication

The communication routines we have seen so far are blocking, meaning the send routine, for example, will not return unless the data has been copied out of the buffer. MPI offers nonblocking communication which is an alternative mechanism that often leads to better performance. The idea is to minimize the time spent to communicate by overlapping communication with computation. A nonblocking **routine (send/recv) start** call initiates the send/receiving operation, but does not complete it. The routine start call will return before the message was copied out of the send buffer. A separate routine complete call is needed to complete the communication, i.e., to verify that the data has been copied/received in/out of the send/receive buffer.

Thus the communication process is separated into two phases: the **initialization** and the **completion**. The idea is then **to do some work** between these two phases. The nonblocking is also useful to avoid deadlock situations. Figures 8 and 9 illustrate the nonblocking send and receive. Here the send and the receive are initiated with *MPI_Isend()* and *MPI_Irecv()* and complete with the *MPI_Wait()*.

Nonblocking send start calls can use the same four modes as blocking sends: standard, buffered, synchronous and ready. These carry the same meaning.

5.1 Communication Objects

Nonblocking communications use opaque request objects to identify communication operations and match the operation that initiates the communication with the

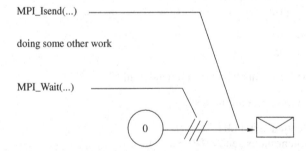

Fig. 8 Illustration of a nonblocking send

MPI_Irecv(...)

doing some other work

MPI_Wait(...)

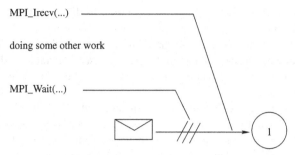

Fig. 9 Illustration of a nonblocking receive

operation that terminates it. These are system objects that are accessed via a handle. A request object identifies various properties of a communication operation, such as the send mode, the communication buffer that is associated with it, its context, the tag and destination arguments to be used for a send, or the tag and source arguments to be used for a receive. In addition, this object stores information about the status of the pending communication operation.

5.2 Nonblocking Send

The names of nonblocking routines start with *MPI_I*. Their syntax (below) is very similar to the blocking routines we have seen earlier. The difference between the blocking and the nonblocking receive is that the nonblocking does not have a status output argument. We will see that this information is available in the completion routines.

```
MPI_ISEND(buf, count, datatype, dest, tag,
          comm, request)

MPI_IRECV(buf, count, datatype, source, tag,
          comm, request)
```

5.3 Communication Completion

To complete a nonblocking communication the functions *MPI_WAIT* and *MPI_TEST* are used. With *MPI_Wait* one waits for the completion of a nonblocking routine. In this case *MPI_Wait* will return only when the operation is completed. *MPI_Test* tests for the completion of a nonblocking routine and will normally always return.

The completion of send and the completion of receive have different meanings. The completion of a send operation indicates that the sender is now free to update the locations in the send buffer (the send operation itself leaves the content of the send buffer unchanged). It does not indicate that the message has been received; rather, it may have been buffered by the communication subsystem. Or, if a synchronous mode send was used, the completion of the send operation indicates that a matching receive was initiated, and that the message will eventually be received by the matching receive.

The completion of a receive operation indicates that the receive buffer contains the received message, that the receiver is now free to access it, and that the status object is set. It does not indicate that the matching send operation has completed (but indicates, of course, that the send was initiated). For more details about the semantics of communication completion we refer the reader to the following documentations [1–4].

Below is a simple example of a program using nonblocking send and receive.

```
MPI_COMM_RANK(comm, rank, ierr)
IF(rank == 0) THEN
   MPI_ISEND(a(1), 10, MPI_REAL, 1, tag,
            comm, request, ierr)
      !!!do some computation to mask latency!!!
   MPI_WAIT(request, status, ierr)
ELSE
   MPI_IRECV(a(1), 15, MPI_REAL, 0, tag,
            comm, request, ierr)
      !!!do some computation to mask latency!!!
   MPI_WAIT(request, status, ierr)
END IF
```

5.3.1 Nonblocking Communication: Discussion

Send and receive can be blocking or nonblocking, a blocking send can be used with a nonblocking receive and vice-versa. A nonblocking send or receive follows directly by a *MPI_Wait* is equivalent to a blocking send or receive. The completion of one message can be achieved using *MPI_Waitany* or *MPI_Testany*. One can also wait or test for the completion of all messages using *MPI_Waitall* or *MPI_Testall*. To wait or test for a the completion of as many messages as possible one use *MPI_Waitsome* or *MPI_Testsome*. The routine *MPI_Request_free(Request)* can be used to deallocate a request handle and should be used to deallocate request handle after the completion of a nonblocking communication.

5.4 Exercise: Rotating Information Around a Ring

In this exercise we suppose that processors are virtually arranged in a ring, see Fig. 10.

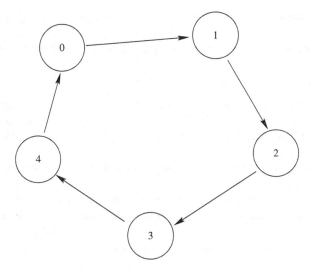

Fig. 10 Five processes logically arranged in a ring: the left neighbor of P_0 is P_4 and its right neighbor is P_1

1. Each processor stores its rank in *MPI_COMM_World* into an integer variable *send_buf*.
2. Each process passes it on to its right neighbor.
3. Each processor calculates the sum of all values.
4. Keep passing it around the ring until the value is back where it started, i.e., each process calculates the sum of all ranks.
5. Use nonblocking send to avoid deadlocks and to verify correctness, because blocking synchronous send will cause a deadlock.

5.5 Extra Exercise

1. Modify your program to experiment with the various communication modes and the blocking and nonblocking forms of point-to-point communications.
2. Modify the above program in order to estimate the time taken by a message to travel between adjacent processes along the ring. What happens to your timings when you vary the number of processes in the ring? Do the new timings agree with those you made with the ping-pong program?

6 Collective Communication

Collective communication unlike point-to-point communication may involve more than two processes at a given time. They are higher level communication routines that may allow optimized internal implementation, e.g. use of tree-based algorithms.

Collective communication is a collective action over a communicator, i.e., all processes in that communicator **must** call the collective routine. However, some of the routine arguments are not significant for some processes and can be specified as dummy values.

Some characteristics of collective communication include:

- Collective communications cannot interfere with point-to-point communications and vice versa. Collective and point-to-point communications are transparent to one another. For example, a collective communication cannot be picked up by a point-to-point receive.
- A collective communication may or may not synchronize the processes involved.
- As usual, completion implies the buffer can be used or reused. However, there is no such thing as a nonblocking collective communication in MPI.
- There are no tags and receive buffers must have exactly the same size as send buffers.
- Datatypes must match between send and receive.

6.1 Barrier Synchronization

```
MPI_BARRIER (COMM)
```

This routine is used when one needs to synchronize processes in a group at certain point in the code. *MPI_barrier(COMM)* blocks the caller until all processes member of the communicator COMM have called it. The call returns at any process only after all group members have entered the call. This routine is very expensive and is normally never needed because synchronization can be done automatically by data communication. Make sure you really need it!

6.2 Broadcast

```
MPI_BCAST(BUFFER, COUNT, DATATYPE, ROOT, COMM)
```

MPI_BCAST broadcasts a message from the process with rank root to all processes of the group, itself included. On return the value of BUFFER will be copied in the buffer used by the other processes to call this routine. This routine should be called by all the processes member of the group associated with the communication COMM. Figure 11 illustrates the broadcast operation.

6.3 Gather

```
MPI_GATHER( sendbuf, sendcount, sendtype,
            recvbuf, recvcount, recvtype, root, comm)
```

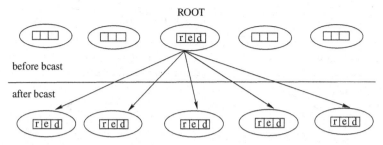

Fig. 11 Illustration of the bcast operation; here the array red is copied from process with rank ROOT to other processes

MPI_Gather gathers data in processors member of the group associated to the communicator into a processor with rank root. See Fig. 12. The outcome is the same as having all processes, the root included, sending the data in sendbuf to the root and the root receiving. Recvcount indicates the number of items received from each process by the root, not the total number of items it receive. All arguments to the function are significant on process root, while on other processes, only arguments sendbuf, sendcount, sendtype, root, comm are significant. The arguments root and comm must have identical values on all processes.

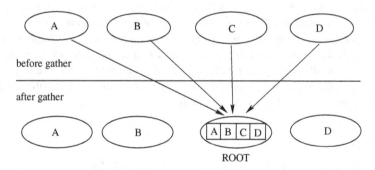

Fig. 12 Illustration of the gather operation

6.4 Scatter

```
MPI_SCATTER(sendbuf, sendcount, sendtype,
            recvbuf, recvcount, recvtype, root, comm)
```

MPI_SCATTER is the inverse operation to *MPI_GATHER*. The outcome is as if the root executed n send operations and each process executed a receive. See Fig. 13.

Variants of these collective communication routines exist. They offer more flexibilities as far as the root and the number of items to receive (for example)

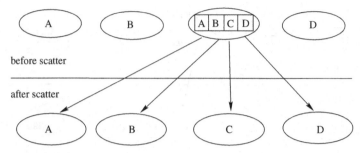

Fig. 13 Illustration of the scatter operation

are concerned for example. One can cite: *MPI_ALLGATHER, MPI_ALLTOALL, MPI_SCATTERV, MPI_GATHERV, MPI_ALLGATHERV, MPI_ALLTOALLV.* See the standard [4] for more details.

6.5 Global Reduction Operations

Instead of merely gathering data from all processors to one processor, one may like to perform an operation on these gathered data. MPI provides a mechanism called **global reduction** to achieve this. For example one can decide to sum the values in a particular buffer on all processors and store the resulting sum in one processor. This type of operations can be done by calling the routine *MPI_Reduce*. Many variants of global reduction functions exist: a reduce that returns the result of the reduction at one node, an all-reduce that returns this result at all nodes and a scan parallel prefix operation. In addition, a reduce-scatter operation combines the functionality of a reduce operation and of a scatter operation.

6.5.1 Reduce

```
MPI_REDUCE( sendbuf, recvbuf, count, datatype, op,
            root, comm)
```

MPI_REDUCE combines the elements provided in the input buffer of each process in the group – using the operation *op* – and returns the combined value into the output buffer of the process with rank *root*. The input buffer is defined by the arguments sendbuf, count and datatype, and the output buffer is defined by the arguments recvbuf, count and datatype; both need to have the same number of elements of the same type. Figure 14 gives an illustration of the reduce operation. Table 4 gives the list of predefined operations. Note that you can also define your own operation; this is usually the case when you are dealing with complex datatypes; see the standard for more details.

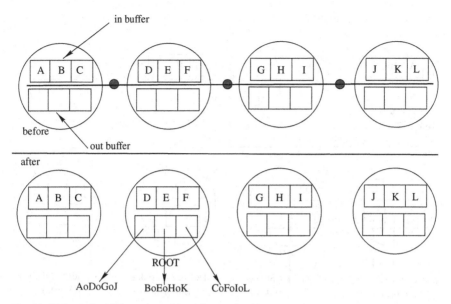

Fig. 14 (Color online) Illustration of the reduce operation

Table 4 Predefined global reduction operations

Predefined reduction operation	Meaning
MPI_MAX	Maximum
MPI_MIN	Minimum
MPI_SUM	Sum
MPI_PROD	Product
MPI_LAND	Logical and
MPI_BAND	Bit-wise and
MPI_LOR	Logical or
MPI_BOR	Bit-wise or
MPI_LXOR	Logical exclusive or
MPI_BXOR	Bit-wise exclusive or
MPI_MAXLOC	Maximum and the location of the maximum
MPI_MINLOC	Minimum and the location of the minimum

6.5.2 All-Reduce

```
MPI_ALLREDUCE( sendbuf, recvbuf, count, datatype,
               op, comm)
```

The same as *MPI_REDUCE* except that the result appears in the receive buffer of all the group members, see Fig. 15. MPI includes variants of each of the reduce operations where the result is returned to all processes in the group.

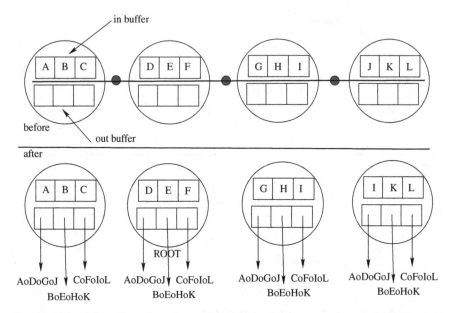

Fig. 15 (Color online) Illustration of the all-reduce operation: the same result in the buffer of all the processes

6.5.3 Scan

```
MPI_SCAN( sendbuf, recvbuf, count,
          datatype, op, comm )
```

MPI_SCAN is used to perform a prefix reduction on data distributed across the group. The operation returns, in the receive buffer of the process with rank i, the reduction of the values in the send buffers of processes with ranks $0, \ldots, i$ (inclusive). Figure 16 gives an illustration of the prefix scan operation.

6.5.4 Exercise: Global Reduction

- Global reduction
  Rewrite the pass-around-the-ring program to use the MPI global reduction to perform the global sum of all ranks of the processes in the ring, i.e., the pass-around-the-ring communication loop must be totally substituted by one call to the MPI collective reduction routine.
- Global scan

  – Rewrite the last program so that each process computes a partial sum.
  – Rewrite it in a way that each process prints out its partial result in the correct order:

  rank=0 → sum=0
  rank=1 → sum=1

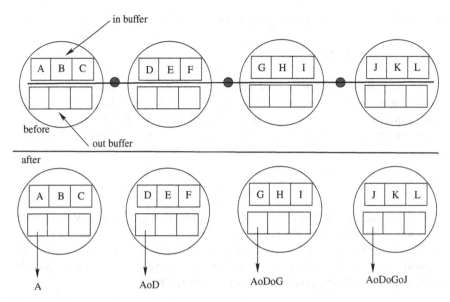

Fig. 16 (Color online) Illustration of the scan operation

rank=2 $\to$ sum=3
rank=3 $\to$ sum=6
rank=4 $\to$ sum=10

7 Communicators, Groups and Virtual Topologies

7.1 Communicator and Groups

We have seen in the previous sections that by default all processors are members of the *MPI_COMM_WORLD* communicator. MPI offers the possibility to create subgroups of processors and associate to these subgroups communicators. There exists also number of functions to manipulate created groups.

7.2 MPI_Comm_Group, MPI_Group_Incl, MPI_Group_Excl, MPI_Comm_Create, Etc.

The simplest way to create groups of processors is to use the MPI routines MPI_Comm_group, MPI_Group_incl, MPI_Group_excl and MPI_Comm_create.

```
MPI_COMM_GROUP(comm, group)
```

MPI_COMM_GROUP returns in group a handle to the group of comm.

```
MPI_GROUP_INCL(group, n, ranks, newgroup)
```

The function *MPI_GROUP_INCL* creates a group newgroup that consists of the n processes in group with ranks rank[0],..., rank[n − 1]; the process with rank i in newgroup is the process with rank ranks[i] in the group. Each of the n elements of ranks must be a valid rank in group and all elements must be distinct, or else the program is erroneous.

```
MPI_GROUP_EXCL(group, n, ranks, newgroup)
```

The function *MPI_GROUP_EXCL* creates a group of processes newgroup that is obtained by deleting from group those processes with ranks ranks[0] ,..., ranks[n − 1].The ordering of processes in newgroup is identical to the ordering in group. Each of the n elements of ranks must be a valid rank in group and all elements must be distinct; otherwise, the program is erroneous.

```
MPI_COMM_CREATE(comm, group, newcomm)
```

This function creates a new communicator newcomm with communication group defined by group and a new context. We can use the above-described routines in the following pseudo code to create two groups of odd and even processors rank number like the one in Fig. 17.

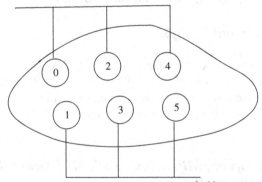

group of even processor rank numbers

group of odd processor rank numbers

Fig. 17 MPI allows the programmer to create group of processors using a communicator

```
...
Odd_ranks={1, 3, 5}, Even_ranks={0, 2, 4}
MPI_comm_group(MPI_COMM_WORLD, Old_group)
```

```
MPI_Group_incl(Old_group, 3, Odd_ranks, Odd_group)
MPI_Group_incl(Old_group, 3, Even_ranks, Even_group)
MPI_Comm_create(MPI_COMM_WORLD, Odd_group, Odd_Comm )
MPI_Comm_create(MPI_COMM_WORLD, Even_group, Even_Comm)
...
```

MPI offers a number of routines for groups and communicators management. We refer the reader to this documentation [4] for more details.

7.3 Virtual Topologies

A virtual topology is a mechanism for naming the processes in a communicator in a way that fits the communication pattern better. It is therefore a convenient process naming that simplifies code witring. It can also allow MPI to optimize communications. Virtual topologies are mostly used when one wants to map the physical structure of its problem into process ranking. For example suppose you are dealing with large multi-dimensional array in your program. If your array is a three-dimensional array, then it would be convenient to arrange your processors logically in a three-dimensional way. Therefore moving from the linear ranking $(0, \ldots, \text{size} - 1)$ to a somewhat more complex ranking. This can be done by integers arithmetic but MPI provides number of routines to handle complex mappings more conveniently. MPI provides **graph** and **Cartesian** topologies. Cartesian topology is suitable for grid topologies. MPI also allows completely general **graph** virtual topologies, in which a process may be "connected" to any number of other processes and the numbering is arbitrary. These are used in a similar way to Cartesian topologies, although of course there is no concept of coordinates. In this course we will concentrate on Cartesian virtual topologies as they are widely used. We refer the reader to the documentation in [4] for details on the graph virtual topology.

7.3.1 Creating a Cartesian Virtual Topology

```
MPI_CART_CREATE(comm old, ndims, dims,
                periods, reorder, comm cart)
```

MPI_CART_CREATE returns a handle to a new communicator to which the Cartesian topology information is attached. Here *ndims* (integer) is the number of dimensions of Cartesian grid and *dims* is an integer array of size ndims specifying the number of processes in each dimension. Figures 18 shows an example of a Cartesian virtual topology. The logical array *periods* of size ndims specifying whether the grid is periodic (true) or not (false) in each dimension, *reorder*, is a boolean input which indicates whether the ranking may be reordered (true) or not (false) (logical). The purpose of reordering the ranks of processors is to choose an optimal mapping of the virtual topology on to the physical machine.

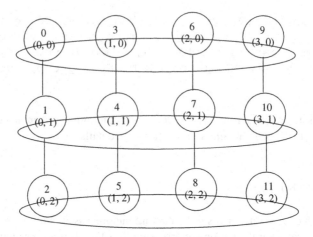

Fig. 18 Example: Graphical representation of a two-dimensional Cartesian topology. Here ndims = 2 dims = (4, 3), periods= (1/.true., 0/.false.), reorder = 0

7.3.2 Cartesian Convenience Function: MPI_DIMS_CREATE

In the example 18 we manually set the dimensions of the Cartesian topology. The function *MPI_DIMS_CREATE* is provided by MPI and allows the user to select a balanced distribution of processes per coordinate direction, depending on the number of processes in the group to be balanced and optional constraints that can be specified by the user.

```
MPI_DIMS_CREATE(nnodes, ndims, dims)
```

MPI_DIMS_CREATE takes the number of nodes in the grid (*nnodes*) and the number of Cartesian dimensions *ndims* and returns in *dims* the *convenient* size of the Cartesian topology. If dims[i] is set to a positive number, the routine will not modify the number of nodes in dimension i; only those entries where dims [i] $= 0$ are modified by the call.

7.3.3 Cartesian Mapping Functions: My Rank, My Coordinates, My Neighbors' Coordinates, Etc.

MPI provides some routines that convert from Cartesian coordinates to the linear ranking and vice versa.

```
MPI_CART_RANK(comm, coords, rank)
```

will return in *rank* the rank of the processor whose Cartesian coordinate are *coords*. The reverse operation

```
MPI_CART_COORDS(comm, rank, maxdims, coords)
```

will return in *coords* the coordinates of the processor whose rank is *rank*.

Another useful Cartesian mapping function is the *MPI_CART_SHIFT*. This function is mostly used in conjunction with *MPI_Send_Recv*. It provides a way to compute the source and the destination needed in *MPI_Send_recv*.

```
MPI_CART_SHIFT(comm, direction, disp,
                rank_source, rank_dest)
```

Here *direction* is the direction in which a shift has to be performed. A positive *displacement* indicates an upward shift while a negative displacement indicates downward shift. On return *rank_source* and *rank_dest* contain the rank of the source and the destination, respectively. Figure 19 shows an example of the shift operation.

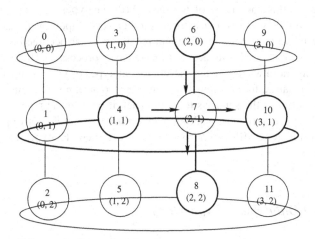

Fig. 19 Illustration of the shift operation. For (dimension, displacement) = (0, +1). In processor 7 we have rank_source = 4 and rank_destination=10. If (dimension, displacement) = (1, +1) we have rank_source=6 and rank_destination=8

7.3.4 Partitioning of Cartesian Structures

If a Cartesian topology has been created with *MPI_CART_CREATE*, the function *MPI_CART_SUB* can be used to partition the communicator group into subgroups that form lower-dimensional Cartesian subgrids, and to build for each subgroup a communicator with the associated subgrid Cartesian topology.

```
MPI_CART_SUB(comm, remain_dims, newcomm)
```

The *i*th entry of logical vector *remain_dims* specifies whether the ith dimension is kept in the subgrid (true) or is dropped (false). This function returns in *newcomm* the communicator containing the subgrid that includes the calling process. Other functions exist to manipulate Cartesian topology; we refer the reader to this documentation [4].

If comm defines a $2 \times 3 \times 4$ grid, and remain_dims = (TRUE, FALSE, TRUE), then *MPI_CART_SUB(comm, remain_dims, newcomm)* will create three communicator-tors each with eight processes in a 2×4 grid.

8 Derived Datatype

So far all the MPI routines we have seen used basic datatypes. Usually basic MPI datatypes correspond to C or Fortran datatypes. However, very often program-mers deal with more complex datatypes that are constructed from existing simple datatypes like integer, real and character. This is for example the case when a pro-gram defines a structure consisting of an array of integer numbers and an array of real numbers. In this case none of the basic MPI datatypes seen up to here can be used to send/receive such a structure. Another very important point to note is that all the data we have been sending and receiving so far occupied a contiguous space in the memory of the sender and the receiver processes. This is obviously not the case any more when you have structure datatypes or when you want to send sub-blocks of a matrix, for example. One solution is to pack noncontiguous data into a contiguous buffer at the sender site and unpack it back at the receiver site. This has the disadvantage of requiring additional memory-to-memory copy operations at both sites. Fortunately, MPI offers routines to create new datatypes from existing basic datatypes.

8.1 Derived Datatypes – Type Maps

Any datatype is specified by its type map, that is a list of the form (Table 5):

Table 5 Derived datatypes – Type maps

Basic datatype 0	Displacement of datatype 0
Basic datatype 1	Displacement of datatype 1
Basic datatype 2	Displacement of datatype 1
. . .	. . .
Basic datatype $n - 1$	Displacement of datatype $n - 1$

A derived datatype is therefore a pointer to a list of entries. Figure 20 shows the memory layout of a struct of one character, two integers and one double precision real number.

In this course we will cover the description of three MPI routines used to create new datatypes, namely *MPI_Type_contiguous*, *MPI_type_vector* and *MPI_type_struct*. We refer the reader to this documentation [4] for other MPI-derived datatypes con-structors.

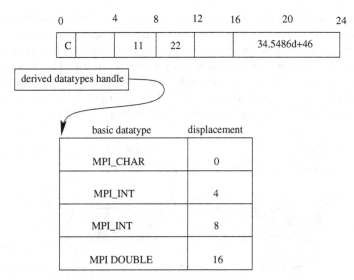

Fig. 20 A derived datatype describes the memory layout of, e.g., structures, common blocks, subarrays and some variables in the memory

8.2 MPI_TYPE_CONTIGUOUS

The simplest datatype constructor is *MPI_TYPE_CONTIGUOUS* which allows replication of a datatype into contiguous locations.

```
MPI_TYPE_CONTIGUOUS(count, oldtype, newtype)
```

The *newtype* is the datatype obtained by concatenating *count* copies of *oldtype*. See Fig. 21.

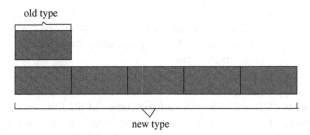

Fig. 21 (Color online) *MPI_type_contiguous* allows the replication of an old datatype to create a new datatype, in this case with count=5

8.3 MPI_TYPE_VECTOR

```
MPI_TYPE_VECTOR( count, blocklength, stride,
                oldtype, newtype)
```

The function *MPI_TYPE_VECTOR* is a more general constructor that allows replication of a datatype into locations that consist of equally spaced blocks. Each block is obtained by concatenating the same number of copies of the old datatype. The spacing between blocks is a multiple of the extent of the old datatype. Here *count* is the number of blocks, *blocklength* is the number of elements in each block and *stride* is the number of elements between the begining of each block. Figure 22 shows an example of construction of a new datatype using *MPI_TYPE_VECTOR*.

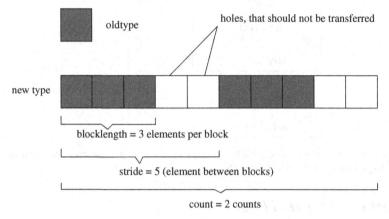

Fig. 22 (Color online) Illustration of a call to *MPI_TYPE_VECTOR* with count = 2, stride = 5 and blocklength = 3

8.4 MPI_TYPE_STRUCT

The most general type constructor is *MPI_TYPE_STRUCT*. It allows each block to consist of replications of different datatypes.

```
MPI_TYPE_STRUCT(count, array_of_blocklengths,
                array_of_displacements, array_of_types,
                newtype)
```

The new datatype *newtype* consists of a list of *count* blocks, where the ith block in the list consists of *array_of_blocklengths[i]* copies of the type *array_of_types[i]*. The displacement of the ith block is in units of bytes and is given by *array_of_displacements[i]*. Figure 23 gives an illustration of *MPI_TYPE_STRUCT*.

8.5 Committing a Datatype

Before a datatype handle is used in message passing communication, it needs to be committed with *MPI_TYPE_COMMIT*. This must be done only once.

```
MPI_TYPE_COMMIT(datatype)
```

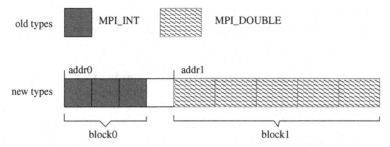

Fig. 23 (Color online) Graphical representation of a datatype created using *MPI_TYPE_STRUCT* for a struct consisting of an array of 3 int and an array of 5 double. Here count = 2 array_of_blocklengths = (3, 5) array_of_displacements = (0, addr1 − addr0) array_of_types = (*MPI_INT, MPI_DOUBLE*)

This operation commits the datatype, that is, the formal description of a communication buffer, not the content of that buffer. Thus, after a datatype has been committed, it can be repeatedly reused to communicate the changing content of a buffer or, indeed, the content of different buffers, with different starting addresses.

8.6 Examples: Sending and Receiving a Row and a Column of Matrices Using MPI Derived Datatypes

In this section we show using C and Fortran codes how to send and receive a column, a row and sub-matrices of a matrix.

8.6.1 Sending and Receiving a Row of a Matrix , C Version

```
. . .
. . .
int nx = 100; // number of rows
int ny = 100; // number of columns
int count_row, blocklength_row, stride_row;

MPI_Datatype        MPI_ROW

count_row       = 1;
blocklength_row = ny;
stride_row      = 1; // arbitrary

MPI_Type_vector(count_row, blocklength_row,
                stride_row, MPI_INT, &MPI_ROW);

 MPI_Type_commit(&MPI_ROW);

// in C elements in the row are contigous
// thus a row can be seen as one block
```

```
// of ny elements. The stride is an arbitrary
// integer number since we have only one
// block

// now the processor with rank 0 sends its first
// row and the 2nd processor receives this row
// and stores it in its sencond row.

int mat[100][100];

if(myrank == 0)
{
MPI_Isend(&mat[0][0], 1, MPI_ROW, 0,
     TAG_BOUNDARY, MPI_COMM_WORLD,
         &request);
}

if(myrank == 1)
{
MPI_Irecv(&mat[1][0], 1, MPI_ROW, 0,
     TAG_BOUNDARY, MPI_COMM_WORLD,
         &request);
}


...
...
```

8.6.2 Sending and Receiving a Row of a Matrix, Fortran Version

```
...
...
integer :: nx, ny
integer :: count_row, blocklength_row, stride_row
nx = 100, ny = 100

count_row       = ny
blocklength_row = 1
stride_row      = nx

integer :: MPI_ROW

call MPI_TYPE_VECTOR(count_row, blocklength_row,
                     stride_row, &MPI_INTEGER,
                     MPI_ROW, IERROR)
call MPI_TYPE_COMMIT(MPI_ROW, IERROR)
```

```
! in Fortran elements in the row are
! NOT contigous thus a row can be seen as ny blocks
! of one elements each. And the number of elements
! between the beginning of each block is
! stride = nx.

integer :: mat(100, 100)

if(myrank == 0) then
call MPI_Isend(mat(1, 1), 1, MPI_ROW, 1,
               TAG_BOUNDARY, MPI_COMM_WORLD,
               request, IERROR)

endif

if(myrank == 1) then

call MPI_Isend(mat(2, 1), 1, MPI_ROW, 1,
               TAG_BOUNDARY, MPI_COMM_WORLD,
               request, IERROR)

endif
```

8.6.3 Sending and Receiving a Column of a Matrix , C Version

```
...
...
int nx = 100; // number of rows
int ny = 100; // number of columns
int count_column, blocklength_column,
int stride_column;

count_row       = nx;
blocklength_row = 1;
stride_row      = ny; // arbitrary

MPI_Datatype      MPI_COL

MPI_Type_vector(count_column, blocklength_column,
               stride_column, MPI_INT, &MPI_COL);

 MPI_Type_commit(&MPI_COL);

// in C elements in the clomun are NOT contigous
// thus a column can be seen as nx blocks
```

```
// of one elements each. And the number of elements
// between the beginning of each block is
// stride = ny.

// now the processor with rank 0 sends its first
// cloumn and the 2nd processor receives this column
// and stores it in its sencond column.

int mat[100][100];

if(myrank == 0)
{
MPI_Isend(&mat[0][0], 1, MPI_COL, 0,
      TAG_BOUNDARY, MPI_COMM_WORLD,
          &request);
}

if(myrank == 1)
{
MPI_Irecv(&mat[0][1], 1, MPI_COL, 0,
      TAG_BOUNDARY, MPI_COMM_WORLD,
          &request);
}
...
...
```

8.6.4 Sending and Receiving a Column of a Matrix , Fortran Version

```
...
...
integer :: nx, ny
integer :: count_column, blocklength_column
integer :: stride_column
nx = 100, ny = 100

count_column       = 1
blocklength_column = nx
stride_column      = 1  ! arbitrary

integer :: MPI_COL

call MPI_TYPE_VECTOR(count_column, blocklength_column,
                     stride_column, MPI_INT,
                     MPI_COL, IERROR)
call MPI_TYPE_COMMIT(MPI_COL, IERROR)
```

```fortran
! in Fortran elements in the clomun are
! contigous thus a column can be seen as 1 block
! of nx elements . And the number of elements
! between the beginning of each block is
! an arbitrary integers (there is only one block)

integer :: mat(100, 100)

if(myrank == 0) then
call MPI_Isend(mat(1, 1), 1, MPI_COL, 1,
                TAG_BOUNDARY, MPI_COMM_WORLD,
                request, IERROR)

endif

if(myrank == 1) then

call MPI_Isend(mat(2, 1), 1, MPI_COL, 1,
                TAG_BOUNDARY, MPI_COMM_WORLD,
                request, IERROR)
endif
...
...
```

8.6.5 Sending and Receiving a Block of a Matrix , C Version

```c
...
...
int nx = 100; // number of rows
int ny = 100; // number of columns
int count_submat, blocklength_submat,
int stride_submat;

//we are sending the leading (nx/4, ny/4)
//!block

count_submat       = nx/4;
blocklength_submat = ny/4;
stride_submat      = ny;

MPI_Datatype       MPI_SUBMAT
MPI_Type_vector(count_submat, blocklength_submat,
                 stride_submat, MPI_INT,
                 MPI_SUBMAT)
MPI_Type_commit(MPI_SUBMAT)
```

```
/* the leading block consists of nx/4 blocks
   of ny/4 elements each and the number of
  between the beginning of each block is
  stride = ny */
int mat[100][100]
if(myrank == 0)
MPI_Isend(mat(1, 1), 1, MPI_SUBMAT, 1,
               TAG_SUBMAT, MPI_COMM_WORLD,
               request)
if(myrank == 1)
MPI_Isend(mat(1, 1), 1, MPI_SUBMAT, 1,
               TAG_SUBMAT, MPI_COMM_WORLD,
               request)
...
...
```

8.6.6 Sending and Receiving a Block of a Matrix , Fortran Version

```
...
...
integer :: nx, ny
integer :: count_submat, blocklength_submat
integer :: stride_submat
nx = 100, ny = 100

!we are sending the leading (nx/4, ny/4)
!block

count_submat       = ny/4
blocklength_submat = nx/4
stride_submat      = nx   ! arbitrary

integer :: MPI_SUBMAT

call MPI_TYPE_VECTOR(count_submat, blocklength_submat,
                     stride_submat, MPI_INTEGER,
                     MPI_SUBMAT, IERROR)
call MPI_TYPE_COMMIT(MPI_SUBMAT, IERROR)

! the leading block consists of ny/4 blocks
! of nx/4 elements each and the number of
! between the beginning of each block is
! stride = nx

integer :: mat(100, 100)
```

```
if(myrank == 0) then
call MPI_Isend(mat(1, 1), 1, MPI_SUBMAT, 1,
               TAG_SUBMAT, MPI_COMM_WORLD,
               request, IERROR)

endif

if(myrank == 1) then

call MPI_Isend(mat(1, 1), 1, MPI_SUBMAT, 1,
               TAG_SUBMAT, MPI_COMM_WORLD,
               request, IERROR)

endif
...
..
```

9 Case Study : The Advection Equation

The advection equation describes the motion of a conserved scalar (represented here by the density ρ) as it is advected by a fixed velocity field v_x, v_y. For simplicity, the velocity field is supposed here to be positive and constant, v_x, $v_y > 0$:

$$\frac{\partial \rho}{\partial t} + v_x \frac{\partial \rho}{\partial x} + v_y \frac{\partial \rho}{\partial y} = 0. \tag{1}$$

The solution of this equation is simply

$$\rho(x, y, t) = \rho \left[x - v_x t, y - v_y t \right]. \tag{2}$$

9.1 Numerical Integration

We divide the two-dimensional plane in $n_x \times n_y$ cells, each of dimension $\Delta x \times \Delta y$, where $\Delta x = L_x/n_x$, and $\Delta y = L_y/n_y$. We represent the conserved variable (ρ) by its values at the discrete set of points $x_i = x_0 + i\Delta x$ and $y_j = y_0 + j\Delta y$, where $i = 0, \ldots, n_x$, $j = 0, \ldots, n_y$, and (x_0, y_0) are the origin's coordinates.

One of the simplest method to determine a stable solution of the advection equation is using the "upwind differencing method" that corresponds to the following discretisation (e.g., Press et al. 1992) :

$$\rho_{i,j}^{n+1} = \rho_{i,j}^n - v_x \frac{\rho_{i,j} - \rho_{i-1,j}}{\Delta x} \Delta t - v_y \frac{\rho_{i,j} - \rho_{i,j-1}}{\Delta y} \Delta t. \tag{3}$$

9.2 Initial Conditions

The dimensions of the computational domain (L_x, L_y) (defined by the parameters XMAX, YMAX) are both equal to 1, while the number of points are indicated by NGRIDX and NGRIDY. The density ρ is fixed initially to 10 inside a circle of radius RC=0.1 positioned at XC=0.5, YC=0.5, and equal to 1 elsewhere. The velocity field is defined by ($v_x = 1$, $v_y = 1$).

10 Overview of the Tasks

In this case study you will learn how to

- Use a master–slave model.
- Perform a domain decomposition and do halo swaps.
- Implement a message passing form of the advection equation.

Task 1: Serial implementation

Write a simple program to implement the above described equation. Increase the size of the grid. Add some timing calls in your program.

Task 2: Parallel implementation

For this part you have to write an MPI version of your serial program. A template program is provided to help you but feel free to implement everything from scratch. As a guide use the provided serial code.

To implement the message-passing version of the code is necessary to

- Generate a Cartesian virtual topology.
- Partition a two-dimensional array between processors.
- Create a master–slave model and map the virtual topology to the two-dimensional array partition.
- Create MPI-derived datatypes to swap halos regions and send back data to the master.

STEP 1: Creating virtual topology

Create a periodic two-dimensional virtual topology. You have to implement a routine **create_cart_topology**() of which a template is provided. The asterisks have to be replaced throughout the program by the appropriate text.

```
MPI Routines required: MPI_DIMS_CREATE,
MPI_CART_CREATE, MPI_CART_COORDS.
```

STEP 2: Initializing local arrays

Each processor has to calculate its own grid size and allocate memory for that grid. If (NGRIDX, NGRIDY) is the size of the global grid then

$$(nx = NGRIDX/dims(1), ny = NGRIDY)/dims(2))$$

is the size of the sub-grid in Fortran and

$$(nx = NGRIDX/dims[0], ny = NGRIDY/dims[1])$$

in C/C++. The array **dims** contents the dimension of the Cartesian topology and is assigned in the routine **create_cart_topology()**. Figure 24 illustrates the data distribution in the domain decomposition approach.

nx = NGRIDX /(number of processors in th X direction)

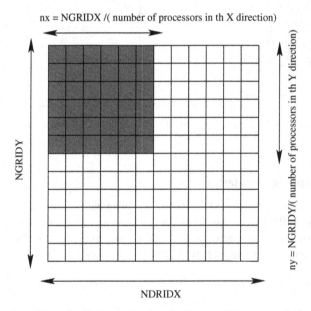

Fig. 24 Illustration of data distribution in the domain decomposition approach. In this case with four processors arranged in a two Cartesian matrix

The next thing to do is to allocate memory for sub-arrays and the global array. The space for the global array is allocated only in one processor: the master processor. In the program the rank of the master processor is 0, ROOT = 0. The routine to modify here is **allocate_dens()**.

Then each processor has to initialize its own local density array **dens**. To do this you must know where the processor lies in relation to the global data space. This can be derived from the Cartesian coordinate (x, y) in the virtual topology in which the processor lies. If (coord[0], coords[1]) is the coordinate of a given processor then we have the following mapping:

$$\text{dens}(i, j) = \text{globaldens}(\text{coords}(0) * nx + i, \text{coords}(1) * ny + j).$$

Use this mapping to initialize the array dens in **init_dens()**. See the Fig. 25.

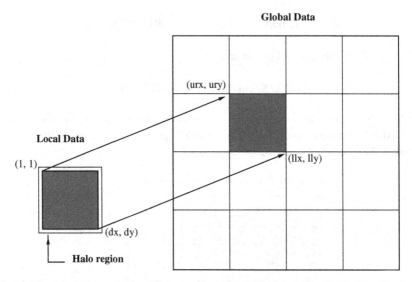

Fig. 25 Mapping between the local data coordinates and the global data distribution. The local array will have a halo region not shown in the above diagram

STEP 3: Swapping boundaries

Here, you have to create derived datatypes and swap halos regions.

First create MPI-derived datatypes for columns and rows. Remember in C elements in the row are contiguous and in Fortran elements in the column are contiguous. The routine to be modified here is **create_mpi_datatypes()**

```
MPI Routines required: MPI_TYPE_VECTOR, MPI_COMMIT.
```

Then each processor has to find its own neighbors to send and receive halos.

```
MPI Routines required: MPI_CART_SHIFT
```

Using the new MPI datatypes and the calculated neighbor processor ranks implement the routine **boundaries_swap()**. Figure 26 illustrates the transferral of rows across the processor domains – the same operation would have to be performed for the columns.

```
MPI Routines required: MPI_ISEND, MPI_IRECV
MPI_WAITALL
```

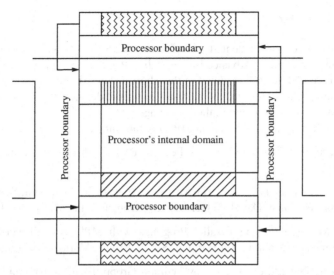

Fig. 26 Exchange of rows across processor boundaries. Exchange of columns is analogous

Step 4 : Sending back local arrays

To visualize the result, all the processors have to send their local array to the master processor. Slave processors have to send local arrays to the master processor. The master processor has to receive the local arrays and store/transfer them into the global array using the mapping describe above, see Fig. 25. Create in **create_mpi_datatypes()** a derived datatype **MPI_SUBMAT** to send back the local array. The routine to modify here is **send_recv_dens()**.

STEP 5 : Writing the main program

Using these routines, write a main program to solve the advection equation.
Test your program with NGRIDX = NGRIDY = 100 and np (number of processor) equals 4;
Increase the size of the grid to 400 on both sides. Use **MPI_WTIME()** to add timing calls in your program. What comment can you make on the scalability of your program?

Congratulations! you wrote a useful MPI program.

STEP 6 : Extra exercise

Generalize the algorithm to make sure that it can deal with any size of grid and/or number of processors.

Further Reading

In this chapter, we have presented a brief overview of the key aspects of the MPI programming techniques. Advanced features like Parallel IO, One Sided Communications etc. could not be presented in this short introduction. However, excellent references are available for readers interested in a more detailed description of MPI. An incomplete list is presented in the following.

Good – introductory level – books on MPI programming are for instance:

– Parallel Programming with MPI, by Peter S. Pacheco, Morgan Kaufmann Publishers, 1997.

– MPI – The complete reference, The MPI Core (vol. 1), by Marc Snir, Steve Otto, Steven Huss-Lederman, David Walker and Jack Dongarra, the MIT press, 1998.

– Writing Message Passing Parallel Programs with MPI, by Neil MacDonald, Elspeth Minty, Joel Malard, Tim Harding, Simon Brown, Mario Antonioletti.

– Using MPI (2nd edition, vol. 1), by William Gropp, Ewing Lusk and Anthony Skjellum, the MIT Press.

while more advanced references are

– MPI – The complete reference, The MPI Core (vol. 2), by Marc Snir, Steve Otto, Steven Huss-Lederman, David Walker and Jack Dongarra, the MIT press. 1998.

– A Fortran Interface to MPI version 1.1 by Micheal Hannecke at http://www.uni-karlruhe.de/~Micheal.Hennecke/#MPI_F90

There are several very good online tutorials available, with extensive examples both in FORTRAN and C. A very partial list include

– Message Passing Interface - MPI at http://www.hlrs.de/organization/sos/par/ services/training/ by HLRS (High Performance Computing Center Stuttgart)

– Intermediate MPI at http://www.osc.edu/press/releases/2002/mpi.shtml, by OSC (Ohio Supercomputer center)

Finally, official copies of the MPI standards (MPI version 1.1 and MPI-2) are available online at http://www.mpi-forum.org

References

1. MacDonald, N., Minty, E., Malard, J., Harding, T., Brown, S., Antonioletti, M., *Writing Message Passing Parallel Programs with MPI*, Edinburgh Parallel Computing Centre, Edinburgh
2. Pacheco, P.S., 1997, *Parallel Programming with MPI*, Morgen Kaufmann Publishers, San Francisco
3. Snir, M., Gropp, W., et al., 1999, *MPI: The Complete Reference*, The MIT Press Cambridge
4. *MPI: A Message-Passing INterface Standard*, The University of Tennessee and Oak Ridge National Laboratory, 2003

An Introduction to Grid Computing Using EGEE

John Walsh, Brian Coghlan, and Stephen Childs

Abstract Grid is an evolving and maturing architecture based on several well-established services, including amongst others, distributed computing, role and group management, distributed data management and Public Key Encryption systems Currently the largest scientific grid infrastructure is Enabling Grids for e-Science (EGEE), comprised of approximately $\sim$250 sites, $\sim$50,000 CPUs and tens of petabytes of storage. Moreover, EGEE covers a large variety of scientific disciplines including Astrophysics. The scope of this work is to provide the keen astrophysicist with an introductory overview of the motivations for using Grid, and of the core production EGEE services and its supporting software and/or middleware (known by the name gLite). We present an overview of the available set of commands, tools and portals as used within these Grid communities. In addition, we present the current scheme for supporting MPI programs on these Grids.

1 Introduction

Grid is an evolving and maturing architecture based on several well-established services, including amongst others, distributed computing, role and group management, distributed data management and Public Key Encryption systems [1]. Currently the largest scientific grid infrastructure is Enabling Grids for e-Science (EGEE) [2], comprised of $\sim$250 sites, $\sim$50,000 CPUs and tens of petabytes of storage. Moreover, EGEE covers a large variety of scientific disciplines, including High

J. Walsh (✉)
School of Computer Science and Statistics, Trinity College Dublin, Dublin, Ireland,
John.Walsh@cs.tcd.ie

B. Coghlan
School of Computer Science and Statistics, Trinity College Dublin, Dublin, Ireland,
coghlan@cs.tcd.ie

S. Childs
School of Computer Science and Statistics, Trinity College Dublin, Dublin, Ireland,
Stephen.Childs@cs.tcd.ie

Walsh, J. et al.: *An Introduction to Grid Computing Using EGEE.* Lect. Notes Phys. **791**, 47–80 (2009)
DOI 10.1007/978-3-642-03370-4_2

Energy Physics (HEP), Geophysics, Astrophysics and Biomedical Computing and these disciplines all have significantly different resource and security requirements.

The authors, as members of the EGEE, Int.EU.Grid [3] and Grid-Ireland [4] communities, present a basic introduction to Grid Computing with an EGEE-centric coverage of the current architecture, middleware and services in use. The scope of this work is to provide the keen astrophysicist with an introductory overview of the motivations for using Grid, and of the core production EGEE services and its supporting software and/or middleware (known by the name gLite [5]). We present an overview of the available set of commands, tools and portals as used within these Grid communities. In addition, we present the current scheme for supporting MPI programs on these Grids. For a comprehensive treatment on using the EGEE Grid, the reader is referred to the gLite Users Guide [6].

1.1 Basic Grid Concepts

Formally, a (computational/data) Grid is a system that is concerned with the integration, virtualisation and management of services and resources in a distributed, heterogeneous environment that supports collections of users and resources (virtual organisations) across traditional administrative and organisational domains (real organisations) [7]. Grids are conceptually based on the model of the provision and consumption of utility services, such as electricity, water and gas supplies [8]. The consumers/users are mostly concerned about the ease of access to and the quality of the provided services. Only very rarely do they want to know about the gory details of how these services are provided to them via the underlying infrastructures, such as water pipes and drainage. In the case of Grid Computing, the infrastructure is often referred to as an e-infrastructure [9]. Grids must be flexible, use open protocols, be secure and must allow co-ordination of resource sharing amongst dynamic collections of individuals, institutions and resources.

1.2 Virtual Organisation Motivations

Academic work, by its nature, often involves rather complex collaborations with people and resources coming from distinct administrative domains. The partners may bring a different set of skills, technology or compute and data resources into the collaboration in order to make it a success. These collaborations may be large, but may be localised, national or may even span international boundaries. Organisation and management of these collaborations give rise to several challenges:

- How do I share my data and resources with someone who does not work locally?
- In large collaborations, it is unlikely that everyone knows each other personally. How do we verify that someone who presents themselves as a member of the collaboration is bone fide?

- How do we ensure that, as a resource provider, we can limit our resources to a well-defined set of known and trusted users?
- Can we scale this in an efficient way, across institutional and, potentially, national boundaries?

Although some of these requirements can be addressed using simple technologies such as secure login shells, secure web servers (e.g. Apache [10] over the https protocol) and secure file transfer with password protection and encryption, it is clear that these do not scale very well. One may manage a small collection of individuals across a small number of domains in such a way, but consider the problem of adding a new member of a virtual organisation to 50 different administrative domains! How do we define or control the access rights that a person or group of people may have on the collection of domain resources? Other issues such as password synchronisation, group management also arise. Grids can address these problems by providing:

- Internationally recognised and secure identity credentials which can be renewed annually or revoked quickly.
- Support for virtual organisations via software and services which respect domain and grid policies.
- Middleware support providing controlled access to resources with additional support for roles and groups, secure access to resources (such as batch systems and storage), subject to the resource providers' policies and single sign-on support (login once, access all matching resources).

Grids based on the gLite middleware supplement the standard Globus-based middleware [11] by providing a way to enumerate and publish information about the resources available (e.g. batch systems), storage usage and storage access interfaces, location, basic site details and the access controls and rights that are applied to site resources. This allows a form of service and resource discovery, permitting a grid user to have resources automatically selected for him/her based on a simple job specification language.

1.3 The International Dimension

Apart from the gLite-based Grids, there are many examples of other Grids with large-scale operations:

- Open Science Grid [12] (USA), an equivalent to EGEE
- TeraGrid [13] (USA), a supercomputing grid
- NAREGI [14] (Japan), a national grid
- National Grid Service [15] (UK), a national grid
- DEISA [16] (EU), a supercomputing grid
- NorduGrid [17] (Nordic Countries), a supra-national regional grid
- Int.EU.Grid [3] (EU), a grid for near-interactive jobs.

As Grid middleware was evolving, the lack of a standardisation body led to divergence in the implementations of some key architectural elements (such as grid storage, the types and formats of information published about each site). The importance of standardisation and interoperability has now been widely recognised and accepted by all of these Grids, and a great deal of effort is being made under the Open Grid Forum's [18] "Grid Interoperability Now" (GIN) [19] initiative.

1.4 The Role of Certificate Authorities

Access to grid resources is based on the establishment of trust relationships between the users and the resource providers. In order for this trust relationship to work in any scalable manner, a key requirement is the establishment of a set of trust authorities (identity providers/certificate authorities) who are mutually recognised by one another. It has been agreed that these identity providers must uphold a common set of agreed standards on establishing the bone fide identity of someone who presents themselves as requiring a Grid identity. The identity providers accept on good faith and with a good degree of confidence that the other identity providers have upheld all the required standards.

Conceptually, the Grid identity providers are similar to passport offices. When a person applies for a passport, he/she must fill out an application form and provide all required details. Additionally, he/she must provide some proof of identity, such as a drivers licence, birth certificate, photographs and signatures. After all the required proof of identity has been provided, the passport office will scrutinise the supplied materials and then accept or reject the application.

In the grid world, application and approval is a multi-staged process:

- The applicant needs to identify their certificate authority from a list of recognised identity providers.[1]
- The identity provider's web site/application form will include a list of registration authorities, where the applicant can physically present themselves, complete with relevant institutional ID.
- The application is typically completed via a web browser form, where the identity provider uses a public key encryption methodology to allow the user to sign their application with a secret key and encrypt the transactions. With the exception of DEISA (and this is being remedied), these grid infrastructures use a common public key encryption system based on the Globus toolkit [20].
- The selected registration authority will require that the applicant presents themselves physically and brings proof of identity with them.
- If the registration authority accepts the identity as bona fide, then the application can be electronically signed and sealed.
- The approved applicant is reminded of the duty of care that is placed upon them in protecting their electronic identity.

---

[1] http://www.eugridpma.org/members/worldmap/

- An electronic credential is then issued, where the applicant's identity cannot be repudiated.
- The credential must be converted by the successful applicant into the X.509 format as used on the Grid.

A great deal of effort is being made by the network and grid communities to simplify the process of acquiring access to the Grid by delegating and federating the identity management to the user's local institution or domain. An interesting example of this is the federated identity management as developed by the Swiss national grid [21].

1.5 The EGEE Grid

The Enabling Grids for e-Science Grid infrastructure is one of the largest and most complex grid infrastructures built to date. It is funded by the European Commission and the partner institutions. EGEE (phase I and II) evolved from the European Data Grid [22] (EDG) and the Large Hadron Collider Grid [23] (LCG) projects. EGEE was comprised of 92 partner institutions from 27 countries and had affiliates in ∼20 other countries. It had over 600 people involved in the overall day-to-day running of the infrastructure. EGEE phase III has now started and is of the same scale. The most salient aims are to

- Provide compute and data storage processing for e-Science;
- Support a large range of scientific communities, each with different requirements, and to encourage new application areas and new user communities;
- Run at production level 24 hours per day, 365 days per year;
- Provide a production-quality infrastructure;
- Develop and maintain innovative grid middleware (gLite) to ease user access and to increase their productivity;
- Run extra testing, certification and pre-production services to ensure overall quality of service;
- Establish grid interoperability with other leading grid projects;
- Encourage the adoption of grid technology in business and commercial domains,
- Provide and ensure top quality administrative and technical management.

At the time of writing, EGEE is composed of ∼250 sites in over 50 different countries, where these countries are partitioned into a set of 10 federations. The current production Grid handles ∼150,000 user jobs per day. The federal structure accommodates the ability to scale the infrastructure management. Each federation has an operations management team which has special responsibility for resolving operational problems of production sites in that region, providing regional helpdesks, and helping new sites to join the Grid by ensuring that candidate sites fulfil a set of well-defined requirements and pass a set of well-defined tests in a pre-production environment. Once a candidate site has satisfied the requirements, it will be allowed to join the production environment.

1.6 Initial Grid Applications

As, historically, the EGEE Grid evolved from the EDG and the LCG Grids, its initial application domains were geared towards supporting the goals and ambitions of the LHC High Energy Physics (HEP) experiments [24] at CERN. EGEE-I was aimed at establishing the move from the experimental and developmental nature of these Grids and to move them into providing a production environment for supporting generic e-Science applications across as wide a range of disciplines as possible. Initially the requirements of the biomedical research community was regarded as sufficiently juxtapose from those of HEP that it was prioritised as a scientific domain that would bring out the best in the provision of grid services. EGEE-I was more than successful in achieving this goal; many other research domains started to use the Grid for production work. Under EGEE-II the applications and scientific domains were widened further to include Astronomy, Astrophysics, Earth Sciences, Computational Chemistry and Nuclear Fusion amongst others.

The EGEE Grid is not intended to simply provide endless raw processing power and storage resources for large scientific applications. Experiments, or virtual organisations, wishing to use the Grid are expected, to some reasonable degree, to contribute resources back to the grid community. Indeed, as the EGEE is not a resource provider per se (the resources are provided by the constituent sites), the VOs may wish to negotiate with the sites to allow them to have access to the sites' resources. The Grid helps to accommodate access to a larger pool of available resources and to accommodate the management of virtual organisations and their data. In principle, the virtual organisations should contribute as much as they use, with the overall effect being a zero-sum game.

2 The EGEE Grid Architecture and Implementation

The EGEE Grid attempts to support the needs of the general scientific community. Its scale requires a significant degree of management at both the project level and at the site and service levels. We shall concentrate here on how the day-to-day services are managed.

At the service level, the EGEE is rather complicated. We have chosen to present a simplified overview of its structure by partitioning it into a set of "Top" or "Higher" level services which help manage the grid infrastructure and its associated VOs and a set of lower level services, such as those provided by the sites to enable grid users to access typical site-level resources such as batch systems and storage systems. See Fig. 1.

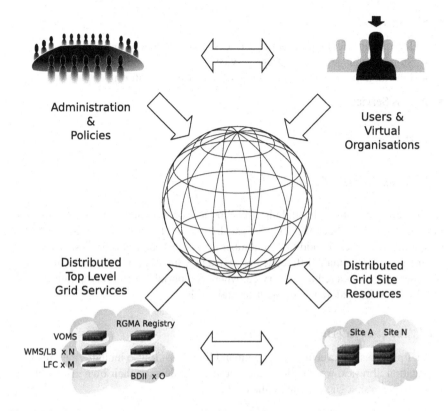

Fig. 1 EGEE actors

2.1 Grid Services

The top-level grid services, which may be centralised or distributed, may be viewed as the set of critical services which enable the Grid to function as an ensemble of

- Administrative Infrastructure

  - Global Grid User Support (GGUS)
  - Grid Operations Centre Database (GOCDB)
  - Service Availabilty Monitoring Service (SAMS) testing
  - Core Infrastructure Centre (CIC)
  - Certificate Authorities (CA)
  - Grid Training

- Core Services

  - Information services
  - Authentication and authorisation
  - Workload management
  - Data management

- Site Services

  - Job management – gatekeeper services, batch systems
  - Storage management – storage services
  - Information – resource and site-BDIIs, R-GMA monitoring

- Access Services

  - User interfaces (UIs)
  - Grid portals

2.2 Accessing the Grid

A grid site may optionally, and almost without exception does, include a grid user interface (UI) ([6], Sect. 3.3.2) – a host that allows the user to interact with the Grid and its services. In addition, a number of web-based portals are also available. Architecturally, the main difference between a user interface and a portal is that a UI is sometimes considered a site service, whereas a portal is definitely a higher-level pan-Grid service and may even span several Grids.

2.2.1 The User Interface

The UI provides a powerful set of command line tools. It also includes a number of application libraries, which allows the grid users to develop their own grid applications. In general, a UI will enable the user to

- List resources suitable to execute a given job,
- Submit jobs for execution,
- Cancel submitted jobs,
- Retrieve the output of finished jobs,
- Show the status of submitted jobs,
- Retrieve extra logging and bookkeeping information for jobs,
- Copy, replicate and delete files from the Grid,
- Discover services and monitor status using the information system,
- Develop grid-enabled applications (APIs/Libraries).

2.2.2 Grid Portals

Portals help address the need for "pervasive access" to the Grid. Access to the Grid and grid services are presented via relatively intuitive and easy-to-use web-based interfaces. VOs are encouraged to actively develop portal applications and their associated web interfaces. Currently, there are a number of portals and portlet servers available, for example:

- Genius [25], based on EnginFrame [26]
- Migrating Desktop [27], from Int.EU.Grid
- P-Grade [28], based on GridSphere [29]

These portals hide the complexities of Grid access mechanisms. They provide certificate, job and data management. Genius is extensively used by EGEE's Gilda [30] test bed for training purposes. Migrating Desktop specifically supports visualisation for near-interactive jobs. P-Grade enables workflow creation, execution, import, export, archival and monitoring, with a built-in graphical workflow editor. P-Grade also provides a single interface to multiple Grids.

2.3 EGEE Sites

An EGEE site must agree to the provision and maintenance of a number of site services. The principal ones are discussed below.

2.3.1 The Gatekeeper

The Gatekeeper ([6], Sect. 3.3.3) is the bridge between the Grid and the resource provider's batch system. The ensemble of the Gatekeeper, batch system and its associated compute nodes is often referred to as a Compute Element (CE) ([6], Sect. 3.3.3). Currently there are a number of different flavours of compute elements supported by EGEE: PBS (OpenPBS [31] and Torque/Maui [32]), LSF [33], Condor [34] and SGE [35]. Compute Nodes (Worker Nodes (WN)) attach to the batch system to run grid jobs. The Site Manager installs middleware and sets up queues for grid access, as well as addtional policies to enforce fair-sharing. The CE publishes information about the status of the batch system, for example:

- Access policy to Grid Queues (which VOs supported at site)
- Number of CPUs available (per VO)
- Expected time to run a job for particular VOs

This information is used by the higher levels of grid middleware to target appropriate jobs to the CE. Once targeted, those jobs are authenticated via grid mechanisms and the grid user is mapped to a local account. An account from a pool of resusable accounts is obtained (this account is cleaned after the job's completion). Resource usage of the local batch system is logged and republished later. Jobs execute on the WNs (scheduled by the batch system), not on the gatekeeper. For better efficiency the VO may install VO-specific software at the site so that this software does not need to be conveyed through the Grid to the CE with each job.

2.3.2 The Storage Element

An EGEE site must provide one or more Storage Elements ([6], Sect. 3.3.4). The storage elements provide a set of standard grid-enabled access methods (protocols) allowing the grid user to access some of the site's storage facilities. Given the scale of the data sets encountered in Grids, data management has a high priority and sophisticated services are provided for creating, using and maintaining data sets as collections (above filesystems) across pools of disks and tape robots. Just as for the CE, information is published about

- Type of Storage Element and protocols supported,
- Supported VOs and their Access Control,
- Available space for VO,
- Duration that data may be kept at site:

  - Temporary,
  - Pinned (retained until further notice) or
  - Permanent.

Site Managers are expected to ensure that data on a grid site is maintained according to agreed policies.

2.3.3 The Site-BDII Service

Each resource at a site, such as the CE and SE, collects and then publishes information about their state on a regular basis. The information is processed and republished in a well-defined format, known as the "Glue Schema" [36]. The role of a site-BDII ([6], Sect. 5.1.5) is to gather this information from a list of known supported services and resources and then republish this in an aggregated form, showing the overall state of the site.

2.3.4 The Site R-GMA Service

As R-GMA acts as a distributed "virtual database", a site's R-GMA server hosts some of that data and will deal with all R-GMA interaction from clients at the site. The server stores data published from the local clients (*primary producers*) and may also collect data from other sites and republish it (*secondary producers*). See Sect. 5.2.1 of the gLite Users Guide for further details.

2.4 Core Grid-Wide and VO-Wide Services

The core Grid and VO services are responsible for the overall management of the user's access to the Grid.

2.4.1 Information Systems and Information Providers

The provision of consistent state information from all of the resources on the Grid is paramount. Currently there are two commonly used information systems used with the EGEE[2] – the Berkeley Database Information Index (BDII) and the Relational Grid Monitoring Architecture (R-GMA) system.

---

[2] gridICE is also commonly used. It allows queries of historical site status data.

GIIS – BDII

The EGEE BDII ([6], Sect. 3.3.5) implements a Grid Information Index Service (GIIS) via a hierarchical system with three levels of *information providers*. As illustrated in Fig. 2, information is "pulled" from the lower to higher levels.

1. Resource-BDII
2. Site-BDII
3. Top-level BDII

At the lowest level is the *resource-BDII*. This service runs on all machines providing one or more grid services (such as a GateKeeper or MyProxy [37]). The resource-BDII will periodically run a set of scripts to determine the current status of that service, and this information is processed and may have extra infomation relating to supported VOs appended. The next level is the *site-BDII*. This periodically aggregates and caches the data provided by the resource-BDIIs. Similar to the resource-BDII, extra data relating to supported VOs and about the site itself will be appended. At the highest level is the *top-level BDII*. The top-level BDII queries the set of site-BDIIs (as determined from a pre-generated list). As the information is stored in an LDAP [38] database, the status of the Grid or services on the Grid can be queried and filtered using standard *ldap* commands. Further details about querying the BDII may be found in Sect. 5.1 of the gLite Users Guide [6].

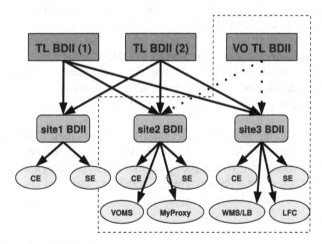

Fig. 2 The hierarchy of informaion

R-GMA

R-GMA [39] is an implementation of the Grid Monitoring Architecture (GMA). It implements a distributed "virtual relational database" which can be manipulated using SQL-like statements.

At an application developer or user level R-GMA is very interesting. As we have seen, the BDII information system deals exclusively with hierarchical site status information. The schema (allowed valid data) is well defined and is "read-only". A user has no ability to modify this data, but may hierarchically query it and filter it. By contrast, R-GMA is relational and with R-GMA the grid user has the ability to create new schemas and tables, insert data with a limited lifetime, relationally query data, subscribe to data publishers (application code that inserts data), etc. R-GMA applications can be built using a number of different languages including C, C++, Python and Java™. A command-line utility can also be used to interact with shell scripts, etc. The R-GMA system is quite extensive and is beyond the scope of this chapter. The reader is refered to Sect. 5.2 of the gLite Users Guide for more details.

2.4.2 Authentication and Authorisation

While authentication (*AuthN*) refers to the process of unrepudiately verifying the identity of a subject (person/host/service), authorisation (*AuthZ*) is the explicit act of denial or approval of access to a service once the subject has correctly authenticated his/her identity. AuthN is supported by a range of facilities as explained in Sect. 1.4. There are two important core AuthN services: CRL services and MyProxy services. AuthZ is supported in EGEE via the VOMS service.

CRL Service
Each Certificate Authority (CA) must publish its certificate revocation list (CRL) via a basic web service so that all dependent services (almost everything) can determine what certificates are no longer valid.

Virtual Organisation Management Service (VOMS)
VOMS [40] is a crucial service in facilitating the organisation of the members of a VO by defining the members' roles and privileges. It is maintained by agreed VO managers. Resource providers have no direct role in the management of a VOMS server.

When a user "logs on" to the Grid, they do so by selecting their VO and optionally one or more of their roles. During the logon, the VOMS server will be contacted and this will yield a short-term VOMS proxy certificate that includes attributes saying what the user's roles and privileges are (an Attribute Certificate (AC)). This dialogue between the user's logon client and the VOMS server is secured via PKI mechanisms.

MyProxy
The MyProxy service is a long-term credential service that a user can employ to

- interact with the Grid from different locations without duplicating their certificate;
- delegate to a portal the power to act on their behalf;
- run jobs that might last longer than the lifetime of a short-lived VOMS proxy certificate.

2.4.3 Workload Management System (WMS)

The WMS [41] is at the heart of managing the user's jobs on the Grid. As such it is a rather complex service which accepts an authorised user's job and assigns it to an appropriate site's CE. It employs a technique known as *match-making* to make the decision to send a job to a chosen CE based on the job's description (expressed using the *Job Description Language*) and other top-level Grid status information. It also manages the staging of small input/output/stderr files – large data should be handled using the data management tools (see Sects. 2.3.2 and 4).

When a job is submitted through the WMS, the state changes are logged in the *Logging and Bookkeeping* (LB) [42] service. This allows the job to be traced fully from the time of submission until after the job's results have been retrieved. The WMS is also capable of implementing some job resilience and can re-submit after some detected failures.

2.4.4 Data Management

There are currently three high-level data management services for the Grid: the LCG File Catalog (*LFC*), the File Transfer Service (*FTS*) and the *AMGA* metadata service. We shall cover the topic of data management in more detail in Sect. 4.

LFC

The LFC [43] lies at the heart of data management and is a critical service for efficient management of files across the grid infrastructure. This service helps the user locate files and their replicas on the Grid. The LFC allows a user to associate an *Alias* (LFN), a user-friendly name, to a file that has been copied onto the Grid using the replica management tools. It will also keep track of all replicas of that file. The LFC implements a hierarchical namespace and allows the user to systematically manage collections of aliases in that namespace. The LFC will publish information into the BDII and this is integrated with the WMS match-making system. This allows jobs to match against sites which store the required files. The use of replicas is a latency-hiding stratagem tantamount to whole-file cacheing.

FTS

The FTS [44] is a new service allowing the user to schedule asynchronous and reliable file replication from one SE to another. It operates similar to a batch system in that the user can submit a "Job" to the FTS. The "Job" has a "state" that can be queried by the user. As such, it is simply an aid to placement of replicas.

AMGA

AMGA [45] is a grid-enabled metadata service. The metadata catalog stores data describing the contents of files registered (using the LFC) on the Grid. The metadata service then allows grid users to search for files that fit a given description. AMGA is an optional service, but is highly recommended where large volumes of data are processed. AMGA is an advanced topic and is beyond the scope of this chapter. The reader is refered to the AMGA users manual [46].

2.5 Administrative Services

2.5.1 GGUS Helpdesk

The *Global Grid Users Support* (GGUS) [47] portal provides a central helpdesk for sites, users and VOs. All problems submitted via GGUS are filtered and distributed to the relevant *Support Units*, i.e. helpdesks or parties. Additionally, GGUS maintains a list of these support units and a list of links to a Frequently Asked Questions (FAQs). Grid operations tickets relating to problematic sites and services are co-ordinated via GGUS. Each EGEE federation provides a team of people to handle operational issues. These teams work on a weekly rotation basis.

2.5.2 Grid Operations Centre DB

The Grid Operations Centre Database (GOCDB) [48] is a service that maintains a list of all sites, site managers, contact details and site status (e.g. the site has not yet been certified to run grid jobs, or the site has been certified and is now a production site). The site manager uses the GOCDB to schedule maintenance periods and to announce these periods to the wider community. The GOCDB may also be queried to list all sites that are currently in production.

2.5.3 Service Availability Monitoring

The EGEE infrastructure is periodically tested via the Service Availability Monitoring system (SAMS) [49]. The test-suite checks if all production services are working correctly. The results of these tests are maintained and visible via a web-interface. In addition, VOs can integrate their own test-suite into SAMS. This allows the VO to test all production sites which support that VO and to use the results of these tests to assemble a list of correctly functional and non-functional production sites. The results can be used to raise tickets with the failing site via GGUS or can be used by VOs who maintain their own top-level BDII to improve the quality of service by guaranteeing that jobs will only go to non-failing sites.

2.5.4 Core Infrastructure Centre

The Core Infrastructure Centre (CIC) [50] provides a portal which provides EGEE management and operational tools; it acts as an entry point for EGEE operational needs to manage available information about VOs and to monitor and ensure grid-wide operations of resources and services.

2.5.5 Grid Training

Grid training and realistic grid training test beds are fundamental to encouraging new users and communities onto the Grid. The EGEE-II NA3 activity coordinates training activities and hosts a repository of training material. EGEE partners are encouraged to run regular training events and to localise the material [51] as needed. In addition, NA3 maintains a list of up-and-coming training events which is available on the EGEE NA3 web site [52]. The *Gilda* [30] training environment gives the curious scientist the oppurtunity to use a dedicated set of training resources via a web portal. Grid-Ireland offers an adaptive e-Learning infrastructure which monitors a user's progress through a set of available courses [53] that can be integrated into discipline-specific syllabii, such as those for pan-institutional Graduate School courses.

3 Computing on the EGEE Grid

In Sect. 1, we argued that one of Grid's many advantages is that it enables researchers to work in distributed collaborations and that it facilitates the secured sharing of resources and results. We now describe how one may perform computational work on the Grid.

The authors do not argue that Grid should be seen as a replacement for high-performance computing (HPC) systems, since Grids and HPC are complementary to one another and both enhance the range and quality of scientific output. HPC systems may be attached to the Grid, as is the case with DEISA or in the case of some EGEE sites where speciality hardware, such as the Myrinet node interconnects, are installed. The vast majority of resources attached to the EGEE Grid are, however, loosely coupled commodity resources without any specialised hardware. In our view, Grid is best suited to high-throughput computing jobs where any of the following conditions are satisfied:

- The jobs have a known medium-term lifetime; however, support for long-running jobs is available via the MyProxy service.
- The jobs are relatively independent, for example Monte Carlo type simulations or Parameter Sweeps.

The Grid can also handle large volume batches of single jobs, complex workflows and MPI jobs (MPI support has been a particular outcome and success of the Int.EU.Grid project). However, cross-site MPI is not widely supported despite the technical success of Int.EU.Grid in providing for this paradigm.

3.1 Running Jobs on the Grid

At the time of writing there is a migration from the existing *gLite 3.0* middleware originally developed by the EU Datagrid (EDG) project and the LCG projects, and further developed by EGEE, towards a new *gLite 3.1* middleware based on web services. This is still taking place across the EGEE infrastructure. The migration involves many changes to how workload management is handled. Consequently it introduces a set of new workload management commands. The examples presented here shall only use gLite 3.0 commands. These are listed in Table 1.

Handling jobs on the Grid is one of its most fundamental operations. In this section we will cover, with examples, jobs that assume the user is a member of a fictitious VO, *myvo*, that illustrate the basics of

- Dealing with grid credentials
- Logging on/off the Grid using VOMS
- The Job Description Language (JDL)
- Matching a job against available resources
- Submitting a job
- Saving the job's global ID
- Querying the job's status
- Cancelling a grid job
- Retrieving the job's output.

Table 1 Current job management commands for EGEE

LCG-2/gLite 3.0	gLite 3.1	Action
edg-job-list-match	glite-wms-job-list-match	Find resources which match job's requirements
edg-job-submit	glite-wms-job-submit	Submit a job to the WMS
edg-job-status	glite-wms-job-status	Get the job status from LB server
edg-job-get-output	glite-wms-job-output	Retrieve the output sandbox of a successful job
edg-job-get-logging-info	glite-wms-job-logging-info	Query logged job information from LB
edg-job-cancel	glite-wms-job-cancel	Cancel a submitted job
	glite-wms-job-delegate-proxy	Delegate a proxy to WMProxy

3.1.1 Handling Grid Credentials

In Sect. 1.4, we discussed the role of Certificate Authorities in establishing the trust relationship between the user and the Grid itself. Once the grid certificate has been approved by the CA, the user must

- Import the .p12 file from CA web site into a web browser
- Save the certificate locally in .p12 format from web browser
- Join a virtual organisation using the VO's secure membership entry page
- Obtain a login account on a UI or portal
- Login to the UI and create a .globus directory on the UI
- Copy the .p12 file to their .globus directory on the UI (using secure copy)
- Unpack the .p12 file to X.509 PEM format
- Set secure permissions on the new files
- chmod 400 userkey.pem to restrict private key read access to only the user
- chmod 644 usercert.pem to allow others to read the public key, but restrict write access to only the user

It is worth stressing that certificates must be kept safe and secure at all times. It is also worth repeating that there are combined efforts within the network and grid communities to greatly reduce the above complexity using federated identity.

3.1.2 Logging Onto/Off the Grid

Single-signon refers to logging onto the Grid, not logging onto the UI. Assuming that the user has already been given an account on a UI and is already logged into it in the usual manner (e.g. via ssh login) and that the user's credentials have been put in place in the user's *.globus* directory, then to signon to the Grid they need to enter:

```
$ voms-proxy-init  --voms myvo
Enter GRID pass phrase:
Your identity: /C=IE/O=Grid-Ireland/OU=tcd.ie/L=RA-TCD/CN=Jo Doe
Creating temporary proxy ................................Done
Contacting  cagraidsvr10.cs.tcd.ie:15013
[/C=IE/O=Grid-Ireland/OU=tcd.ie/L=RA-TCD/CN=host/cagraidsvr10.
cs.tcd.ie] "myvo"
Done
Creating proxy ............................... Done
Your proxy is valid until Tue Jan  8 03:36:29 2008
```

This action creates a *VOMS proxy certificate* – a temporary certificate that is used for grid authentication. The proxy is valid for 12 hours by default. The *voms-proxy-init* command offers a number of extra command-line options which can

- Change the lifetime of the generated proxy (set by the VO managers),
- Use a different location for the certificate rather than the default,
- Assign specific role attributes.

The user may also employ the MyProxy service as a means of obtaining a valid VOMS proxy certificate.

3.1.3 Checking Proxy Certificate Status

As the proxy certificate is copied with the submitted job eventually to a Worker Node, the user must ensure that the proxy does not expire before the job is completed. The proxy certificate status can easily be checked prior to job submission.

```
$ voms-proxy-info
```

This command displays default information about the proxy certificate on standard output. In addition, if the certificate is valid the command will return code 0 to the shell and non-zero otherwise. This allows the command to be used for scripting more complex workflows. Other options to display more specific information about the proxy are to

- check if the proxy exists and is valid (-exists),
- check the remaining proxy lifetime in seconds (-timeleft),
- print all known information about the proxy (-all),
- print the Distinguished Name (DN) of the proxy (-subject).

```
$ voms-proxy-info --all
subject   : /C=IE/O=Grid-Ireland/OU=tcd.ie/L=RA-TCD/CN=Jo Doe/CN=
proxy
issuer    : /C=IE/O=Grid-Ireland/OU=tcd.ie/L=RA-TCD/CN=Jo Doe
identity  : /C=IE/O=Grid-Ireland/OU=tcd.ie/L=RA-TCD/CN=Jo Doe
type      : proxy
strength  : 512 bits
path      : /tmp/x509up_u59999
timeleft  : 11:44:31
=== VO myvo extension information ===
VO        : myvo
subject   : /C=IE/O=Grid-Ireland/OU=tcd.ie/L=RA-TCD/CN=Jo Doe
issuer    : /C=IE/O=Grid-Ireland/OU=tcd.ie/L=RA-TCD/CN=host/cagra
idsvr10.cs.tcd.ie
attribute : /myvo/Role=NULL/Capability=NULL
timeleft  : 11:44:32
```

3.1.4 The Job Description Language (JDL)

The JDL [54, 55] has a simple syntax for describing the properties of the job that the user would like to run on the Grid. The syntax is in the form of key-value pairs, i.e. "Parameter = Value". Pairs can be used to define:

- The Executable,
- Stdin/stdout/stderr Files,
- Input Files,
- Output Files,
- Optional Catalogued Files,
- Optional requirements from known batch systems states,
- Optional memory, number of CPUs, required etc.,
- Optional storage requirements,
- Optional *Rank* requirements expressions.

Rank requirement expressions provide a way to prioritise the selection of site resources. These expressions are highly coupled with information published by the site's BDIIs, so a typical rank requirement expression might be, for example, to match against those sites that use a certain batch system (e.g. PBS) or to match against those sites where the expected waiting time before the job will run is low.

Rank requirements are beyond the scope of this chapter and are covered in the gLite Users Guide Sect. 6 in greater detail.

In the example below, we show how easy it is to specify a basic grid job. The grid job will run the executable script "test.sh". This script will be passed to two arguments. Any errors on stdout while running will be saved to the file "std.err" and stdout will be saved to "std.out". The job will also return an output file called "ResultsOverview.txt". In addition, the job will run on any CE where there are at least 10 free CPUs available and where the host name does not match "cern.ch". The script "test.sh" may be any bash script that one would expect to run on a Worker Node.

```
Executable = "test.sh";
InputSandbox =  {"/home/jodoe/test.sh"};
Arguments = "3.141592653589 data-set-01";
StdOutput = "std.out";
StdError = "std.err";
OutputSandbox = {"ResultsOverview.txt"};
RetryCount = 3;
Rank = {other.GlueCEStateFreeCPUs >= 10};
Requirements  = (!RegExp("cern.ch",other.GlueCEUniqueID));
```

3.1.5 Matching Jobs Against Resources

Before submitting a job to the Grid, the user may wish to see what resources are available. The example below shows, for brevity, a subset of the list of Compute Elements where the grid job can run.

```
$ edg-job-list-match example.jdl

Selected Virtual Organisation name (from proxy certificate
extension): myvo
Connecting to host rb.example.com, port 7772

**********************************************************************
                  COMPUTING ELEMENT IDs LIST
The following CEs matching your job requirements have been found:

                  *CEId*
ce1.example.com:2119/jobmanager-lcgpbs-long
ce2.example.com:2119/jobmanager-pbs-myvo
ce3.example.com:2119/jobmanager-pbs-short
**********************************************************************
```

3.1.6 Submitting a Job

The example below shows how a job may be submitted through a LCG Workload Management System. The output returns a globally unique *Job Identifier* in the form of a URL. This Job ID is then used by subsequent job management commands.

```
$ edg-job-submit example.jdl

Selected Virtual Organisation name (from proxy certificate extens
ion): myvo
Connecting to host rb.example.com, port 7772
Logging to host rb.example.com, port 9002

*********************************************************************
                              JOB SUBMIT OUTCOME
The job has been successfully submitted to the Network Server.
Use edg-job-status command to check job current status. Your job
identifier (edg_jobId) is:

- https://rb.example.com:9000/qEml-pDbzP2NaPs8u3QTPg
*********************************************************************
```

3.1.7 Managing Groups of Jobs

When submitting and managing multiple jobs, job submission command-line options can be used to force the Job ID to be appended to a chosen file. Corresponding job management commands have corresponding command-line options to read in and process Job IDs from a file. Further details can be found in the gLite Users Guide.

3.1.8 Job States

When a job is submitted to the Grid, its progress is tracked via a series of state changes. Under normal execution the job states change as per the progression listed in Table 2.

Table 2 Job status states

	States	Meaning
1.	Submitted	The job was submitted by the user, but not yet processed by the WMS
2.	Waiting	The job has been accepted by the WMS, but has not yet been assigned to a CE
3.	Ready	The job has been assigned to a CE, but the job data has not yet been transferred
4.	Scheduled	The job is queued on the CE's batch system
5a.	Running	The job is running on the CE
5b.	Aborted	The job could not be run
6.	Done	The job has completed on the CE and WN, and an output sandbox has been transfered to the WMS
7.	Cleared	The job's output sandbox has been transferred to the UI

3.1.9 Querying the Jobs Status

The user may also query the default WMS to print the status of all of their jobs. The example below shows the user querying an LCG WMS; however, the gLite 3.1 equivalent has many more command-line options for parsing/selecting a restricted number of jobs matching particular requirements.

```
$ edg-job-status GlobalJob.id
```

```
$ edg-job-status --all
```

The example output below shows the status of a completed job. Note that the "Current Status" indicates that the output of the job has not yet been retrieved.

```
****************************************************************
BOOKKEEPING INFORMATION:

Status info for the Job : https://rb.example.com:9000/3B2
CA1WB5AEgSrQXhgxMOg
Current Status:     Done (Success)
Exit code:          0
Status Reason:      Job terminated successfully
Destination:        ce3.example.com:2119/jobmanager-pbs-short
reached on:         Thu Oct 18 21:26:41 2007
****************************************************************
```

3.1.10 Job Cancellation

It may become necessary to cancel a grid job, for example the job may have become stuck due to a Worker Node at a site failing, or the job details may have been incorrect (e.g. the parameters may have been in error) at submission.

```
$ edg-job-cancel GlobalJob.id
```

3.1.11 Retrieving Job Outputs

To download the job's output sandbox files (as specified in the JDL) to the UI, the following command should suffice:

```
$ edg-job-get-output GlobalJob.id
```

4 Data Management

Data management support offered by EGEE/gLite is one of its greatest strengths. As we have seen, the WMS helps orchestrate the management of jobs in the Grid by copying a small amount of job data from the UI to the CE. Typically this is in the order of 20 MB maximum. This not a large amount of data by today's standards, but does represent a "copy-by-value" chain rather than a more efficient "copy-by-reference" chain. Hence, this system is not intended to copy large volumes of job-related data around the Grid. For better efficiency, specific data management tools and services, independent of the WMS, are needed to enable

- data procurement,
- data distribution/replication,
- data lifetime management.

These tools must be fast and scalable. We examine firstly the problems in dealing with data in a grid environment, and then show how the EGEE/gLite tools help in solving this problem.

4.1 The Storage Problem

In implementing a distributed storage system, a number of issues and challenges arise:

- How do we support copy-by-reference?
- How do we guarantee uniqueness of the names that refer to the data?
- Although the data location may be unique, tracking this is difficult in a changing environment.
- The data often needs to be moved – e.g. it may need to be close to computation facilities.

4.2 The Storage Solution

To solve these problems EGEE/gLite provides namespace services and tools which can "catalog" files on the Grid. Files copied onto the Grid this way are "registered" into a global logical namespace, i.e. when a file is registered in a catalog, it is assigned a unique global identifier. A registered file can be assigned user-friendly "aliases" and these aliases can be used for most data-related operations. The "LFC" catalog service implements a hierarchically structured namespace, allowing the user to store and access data in a location-independent manner, and just refer to it by name during job submission (copy-by-reference). The introduction of *Storage Resource Manager* (SRM) [56] services at many of the sites has enabled sites and VOs to support access controls and security on this registered data.

Note that it is possible to work on the Grid without catalogs by using some of the more basic underlying protocols such as gsiftp [57]. However, in doing so, the user discards the many advantages that the file catalogs bring.

4.3 Grid Files

By "file on the Grid" it is generally accepted that a "file" is

- a data file stored physically on an SE and
- registered in a catalog.

There are several ways for the user to get data into grid storage. The data can be copied onto the Grid using standard data transfer commands, e.g. the data is first copied from the user's desktop to the UI using *scp*, then the user invokes the data management commands provided by the "lcg_utils" suite ([6], Sect. 7.5) – see Sect. 4.6 for further details.

4.4 Write Once/Read Many

Grid facilitates the concept of an optimum location of data that will be processed. This often involves replication of data from the orginal location to those locations that provide the fastest access, best bandwidth, most available space, etc. The "lcg_utils" are designed to deal with this. However, it is important to realise that the original and any replicas must remain consistent. A design decision that helps enforce this consistency was to mandate that grid files are "write once, read many". The implication is that files can be created, read and deleted, but not modified. "Modifications" can be achieved by copying to a new file without registering it and then modifying it. The resulting file can then be copied back to an SE and registered. Naturally, this introduces some logistical overhead as new replicas may need to be copied to other SEs.

4.5 File Namespaces

In reality EGEE supports a number of different file namespaces, so a file can be referred to by a number of names. Each namespace has its own advantages and disadvantages. Table 3 summarises the EGEE namespaces. The "lcg_utils" provides a number of commands to convert from one namespace to another. Examples are shown in Sects. 4.7.8 and 4.7.9.

Table 3 EGEE file namespaces

Namespace	Description
GUID	A globally unique idenitifier Assigned when the file is registered in the LFC Format is consistent, but non-memorable or user-friendly e.g. $guid : 38ed3f60 - c402 - 11d7 - a6b0 - f53ee5a37e1d$
LFN	Logical file name Inituitive user-assigned user-friendly name Name is an *Alias* to a grid file and its replicas Alias is set after the file has been registered Stored in a tree-like structure e.g. */grid/myvo/user4/mylfn_alias*
SURL/PFN	Storage uniform resource location or physical file name Identifies the location of a file as seen by the SE's Storage Manager e.g. *srm://se1.example.com/dpm/example.com/home/myvo/gener* *ated/2008-04-11/file901b5fe6-a16b-4d73-beda-1f80e5f46b93*
TURL	Transport URL Identifies the physical location on the SE's disk Identifies the protocol supported by the SE e.g. *gsiftp://se1.example.com:/storage/myvo/2008-04-11/file2d2521b5-* *4d41-444d-9660-4e7a2f3f1b87.106656.0*

4.6 Data Primitives

The "lcg_utils" data management tools offer an extensive number of commands to manipulate files on the Grid, allowing the user to add, remove and replicate files. Files can be copied between the UI, CE, SE and WN (Table 4).

Table 4 Replica management

Command	Action
lcg-cr	Copies a file to an SE, registers the file in the catalogue (upload)
lcg-cp	Copies a catalogued file to a local destination (e.g. UI or Worker Node) (download)
lcg-del	Deletes one catalogued file (either one replica or all replicas)
lcg-rep	Copies a catalogued file between named SEs, then registers the copy
lcg-gt	Gets the TURL for a given SURL and transfer protocol

In addition, these utilities provide a number of commands to manipulate the file catalog. Some of the most common are tabulated in Table 5.

Table 5 Catalog management

Command	Action
lcg-aa	Adds an alias in the catalogue for a given GUID
lcg-ra	Removes an alias in the catalogue for a given GUID
lcg-la	Lists the aliases for a given LFN, GUID or SURL
lcg-rf	Registers in the catalogue a file residing on an SE
lcg-uf	Unregisters in the catalogue a file residing on an SE
lcg-lr	Lists the replicas for a given LFN, GUID or SURL
lcg-lg	Gets the GUID for a given LFN or SURL

Tracking files on the Grid by reference to an absolute location and path is a complex task. The LFC simplifies this by allowing the user to create easy-to-use and user-friendly *aliases* that can be used by the user to refer to catalogued files. The LFC includes a set of commands for managing these aliases within its hierarchical namespace (Table 6).

Table 6 Namespace management

Command	Action
lfc-ls	Lists files in the LFC namespace
lfc-mkdir	Creates a directory in the LFC namspace
lfc-rename	Renames a file or directory
lfc-rm	Removes a file or directory in the namespace
lfc-chmod	Changes the access permissions on an LFC registered file or directory
lfc-chown	Changes ownership of an LFC registered file or directory
lfc-ln	Creates a symbolic link to an LFC registered file or directory
lfc-getacl	Gets the current access controls on a file or directory
lfc-setacl	Gets the current access controls on a file or directory

4.7 Data Management Examples

We present a simple set of examples of how to copy files onto the Grid, add an alias, replicate the file to another site using that alias, etc. In these examples, we assume that the user (user4) is logged onto the Grid and is a member of the *myvo* virtual organisation. In addition, the examples assume that there is a pair of SEs on the Grid: se1.example.com and se2.example.com.

4.7.1 Listing Files in the LFC Hierarchy

The normal convention when using the LFC is that users will store their registered LFNs under the namespace path of */grid/VONAME/USERDIR/*. One may list files and directories in the LFC namespace as follows:

```
$ lfc-ls /grid/myvo/
SAM
desktop
generated
myvouser1
myvouser2
myvouser3
```

4.7.2 Creating a Subdirectory in the LFC Namespace

The user can create a directory where they can store and manage their aliases (their LFNs).

```
$ lfc-mkdir /grid/myvo/myvouser4
$ lfc-ls /grid/myvo/
SAM
desktop
generated
myvouser1
myvouser2
myvouser3
myvouser4
```

4.7.3 Copying a File from the UI to an SE

The user can copy and register a file on the Grid as follows. The unique global ID of the file on the Grid is returned as output.

```
$ lcg-cr --vo myvo -d se1.example.com \
            file:/home/user4/myfile.txt
guid:2e9f6653-f804-4d66-be76-50d1a0d1defb
```

4.7.4 Adding an Alias

To add a user-defined alias:

```
$ lcg-aa --vo myvo \
            guid:2e9f6653-f804-4d66-be76-50d1a0d1defb \
            lfn:/grid/myvo/user4/myfile_alias.txt

$ lfc-ls /grid/myvo/user4/
myfile_alias.txt
```

Now the LFN lfn:/grid/myvo/user4/myfile_alias.txt can be used to refer to the file. A listing of the directory shows that the alias is added.

4.7.5 Listing Replicas Using the Alias

The user can list a replica of a given LFN as follows. The output shows the SURL format of the file's location.

```
$ lcg-lr lfn:/grid/myvo/user4/myfile_alias.txt
srm://se1.example.com/dpm/example.com/home/myvo/generated
    /2008-04-11/file901b5fe6-a16b-4d73-beda-1f80e5f46b93
```

4.7.6 Replicating the File

The user can now copy the file to another location on the Grid and then list the replicas. Usually a suitable location for such files can be determined by using the client-side information tools. The "lcg-rep" command produces no output unless an error were to occur.

```
$ lcg-rep -d se2.example.com --vo myvo \
        lfn:/grid/myvo/user4/myfile_alias.txt

$ lcg-lr lfn:/grid/myvo/user4/myfile_alias.txt
srm://se1.example.com/dpm/example.com/home/myvo/generated
    /2008-04-11/file901b5fe6-a16b-4d73-beda-1f80e5f46b93
srm://se2.example.com/dpm/example.com/home/myvo/generated
    /2008-04-11/file5def7494-5fa9-41d1-97cb-a152744f48b2
```

4.7.7 Renaming an Alias

The user may wish to rename an alias, for example, to indicate that the data has been processed. In this simple example, the user renames the alias and then lists the replicas. As can be seen, the replica SURLs have not changed.

```
$ lfc-rename /grid/myvo/user4/myfile_alias.txt \
            /grid/myvo/user4/myfile_new_alias.txt
srm://se1.example.com/dpm/example.com/home/myvo/generated
    /2008-04-11/file901b5fe6-a16b-4d73-beda-1f80e5f46b93
srm://se2.example.com/dpm/example.com/home/myvo/generated
    /2008-04-11/file5def7494-5fa9-41d1-97cb-a152744f48b2
```

4.7.8 Retrieving the GUID

In this example, the user knows the SURL, but would like to determine its GUID. The command also works with LFNs. Users may want to do this, for example, if they decide to remove an alias in the catalog but want to keep track of the file for further processing later.

```
$ lcg-lg --vo myvo srm://se1.example.com/dpm/example.com/home
    /myvo/generated/2008-04-11/file901b5fe6-a16b-4d73-beda-1
    f80e5f46b93
guid:2e9f6653-f804-4d66-be76-50d1a0d1defb
```

4.7.9 Retrieving the TURL

One may wish to determine the TURL for a grid file so that it may be manipulated with lower level data management commands such as *globus-url-copy*. Note that the transport protocol needs to be specified in this command.

```
$ lcg-gt srm://se1.example.com/dpm/example.com/home/myvo/
      generated/2008-04-11/file2d2521b5-4d41-444d-9660-4
      e7a2f3f1b87 gsiftp
gsiftp://gridstore.cs.tcd.ie/gridstore.cs.tcd.ie:/storage/
      cosmo/2008-04-11/file2d2521b5-4d41-444d-9660-4e7a2f3f1b87
      .106656.0
106660
0
```

4.8 Deleting a File

In this example, the user removes a file from the Grid and removes it from the LFC. The example also shows that trying to subsequently manipulate the file generates errors:

```
$ lcg-del --vo myvo \
      -a guid:2e9f6653-f804-4d66-be76-50d1a0d1defb

$ lcg-lr lfn:/grid/myvo/user4/myfile_new_alias.txt
lfc-host.example.com: /grid/myvo/user4/myfile_new_alias.txt:
   No such file or directory
lcg_lr: No such file or directory
```

5 Putting it All Together

Managing multiple grid jobs is often viewed as a difficult process. A single job may be one of many tasks from a much larger and complex workflow. Breaking down this work into individual jobs, gathering input, submitting jobs and gathering the output in the correct order requires time, effort and planning.

Portals, such as P-Grade (which has a workflow editor), can help manage this and the introduction and support of workflows based around Directed Acyclic Graphs [55] in gLite and its JDL can certainly ease this problem.

However, as an aid to understanding the handling of a relatively minor sets of tasks, we encourage the reader to explore the following example that encompasses a complete workflow – breaking a task into smaller tasks, generating a set of JDL scripts, submitting a set of jobs, waiting for all jobs to finish and then gathering and processing the collection of job outputs. The example was developed by Kathryn Cassidy for a Grid Tutorial Workshop held in Trinity College Dublin in March 2006, and is freely and publicly available.[3] Although the example was developed for the LCG-2 infrastructure, it can be easily modified to use the gLite 3.1 WMS. The example is comprised of two scripts:

1. submit-dictionary.sh – the workflow script executed by the user
2. echoword.sh – the job that is executed on the Worker Node

The pseudo-code algorithm for the workflow is presented as follows:

```
Check that there is a single command-line argument

if number of command-line arguments is not 1
then
 Print Usage Text
 exit with error
else
 Print number of jobs to submit
endif

set values for location of dictonary loccation,
    JDL job file location and the result file.

for index = 1 to value of command-line argument
  Let RAND be random value between 1 and 45427
  Select the word from line number RAND from dictionary
  Create a JDL file with the index and the word set as
      arguments
  Submit the JDL to WMS and save the Job ID to a file
end for

while jobs are still running
  Check all the job status status changes to one of Done,
      Cancelled, Aborted, Cleared
  if all jobs in one of these status then
  All Jobs are finished
  Exit loop
  end if
  Sleep for a polling interval of 30 seconds
end while

Retrieve the output files for all jobs
Concatenate all outputs into a single Results file.
```

---

[3] http://www.grid.ie/grid-event-2006/march-course/code/dictionary.tgz

6 MPI Support in EGEE

Initially, the support for MPI on the EGEE Grid had been rather weak, whereas MPI was always well supported on the Int.EU.Grid infrastructure. An *EGEE technical working group* discovered [58] that support in the WMS was not flexibile enough, that support for configuring sites for MPI was lacking, that relatively few sites supported it by default, that MPI service discovery was not sufficient due to limitations in the Glue Schema and that many of the sites in EGEE were and are physics-orientated and in some cases did not have any experience in supporting MPI. These issues were not major technical problems, but rather small issues with a big impact on a large MPI user base and their overall experience of the Grid. Indeed, many of the non-HEP communities such as Earth Sciences, Nuclear Fusion, Astrophysics and Computation Chemistry have a major dependence on MPI. These problems have been solved and the solution is gradually being deployed at the EGEE sites (at the time of writing).

6.1 Running MPI on the Grid

In Sect. 3.1.4, we saw a simple example of a grid job expressed using the JDL. The sample code below shows how to grid-enable MPI code using a very similar framework. Further information, including the required wrapper scripts, can be found on the EGEE MPI web page.[4]

```
1 JobType = "MPICH";
2 NodeNumber = 8;
3 Executable = "mpi-start-wrapper.sh";
4 Arguments = "mpi-test OPENMPI";
5 InputSandbox = {"mpi-start-wrapper.sh","mpi-hooks.sh",
6                 "mpi-test.c"};
7 Requirements =
8   Member("MPI-START", other.
        GlueHostApplicationSoftwareRunTimeEnvironment)
9   && Member("OPENMPI", other.
        GlueHostApplicationSoftwareRunTimeEnvironment)
```

Line 1 of the JDL shows that MPI jobs must be submitted with a JobType of *MPICH*. Line 2 sets the number of CPUs or cores required for the job. Most importantly, in line 3, we see that the executable is set to mpi-start-wrapper.sh. This script handles the setting up of the MPI environment and the execution of the real MPI executable as determined by the arguments set in Line 4. The wrapper script, the MPI code and a special set of shell functions (job hooks) are passed in with the input files (lines 5 and 6). The role of these hooks are explained below.

---

[4] http://egee-docs.web.cern.ch/egee-docs/uig/development/uc-mpi-jobs_2.html

Finally, a requirement should be specified (lines 7–9) that selects only sites that have installed the required MPI support scripts and that support the desired version of MPI (OPENMPI in this case). Additional JDL requirements, such as output files, can be added as wished.

6.1.1 mpi-start-wrapper.sh

mpi-start-wrapper.sh is the JDL executable and the user's script that aids the execution of the user's MPI code. It is supplemented by an additional call to a site-installed script defined by the variable **$I2G_MPI_START** which sets up the correct MPI environment and performs the actual execution of the MPI code. It is the site manager's responsibility to install this latter script and to define the environment variable properly.

```
# Setup for mpi-start.
export I2G_MPI_APP=$MY_EXECUTABLE
export I2G_MPI_TYPE=$MPI_FLAVOUR
export I2G_MPI_PRE_RUN_HOOK=mpi-hooks.sh
export I2G_MPI_POST_RUN_HOOK=mpi-hooks.sh
# Invoke mpi-start.
$I2G_MPI_START
```

6.1.2 mpi-hooks.sh

The role of the mpi-hooks.sh script is to provide a user-controlled way of executing pre-processing tasks before the actual MPI program runs and post-processing tasks after the MPI code has executed. In the example below, the pre_run_hook compiles C code using mpicc. Indeed, it is recommended that the user applies the pre_run_hook so that the MPI code is tailored exactly to the site's runtime environment. Similarly, the post_run_hook can be defined to allow the user to process the completed MPI job's output data or to copy and register large files to the Grid, etc.

```
pre_run_hook () {
mpicc -o ${I2G_MPI_APP} ${I2G_MPI_APP}.c
}
```

7 Grid-Secured Applications

The establishment of globally recognised credential services has enabled the scientific community to develop large distributed collaborations. These may take advantage of the underlying PKI infrastructure to grid-secure many applications requiring authentication and authorisation. A very small sample of these and how they are used is described below:

GridSite: "Grid Security for the Web" [59] is an Apache Web server module that allows a grid user's X.509 credential to be used to interact with a web service. A typical use is to allow an authorised user to add or edit web pages on a remote GridSite-enabled web server, upload files to the web server or permit/deny authorised users access to web pages or services. This is a very widely used enhancement to web sites for grid collaboration.

MySQL: Authentication and authorisation to databases can now be established using X.509 credentials [60].

GSI-OpenSSH: A modified openssh can use a grid user's X.509 certificate or proxy certificate to enable secure shell access to remote computers [61].

8 Conclusion

Grids have become more pervasive as a technological means for solving medium to large-scale scientific problems and enabling distributed collaborations to form, operate and coordinate themselves in an efficient manner. The establishment of globally recognised identity credentials and the move to federated identity management will ease access to grid infrastructures and e-infrastructures.

However, despite the rather noble goals set by many of the grid infrastructures, there remain weak points that still need to be addressed. It is as yet unclear whether Grids meet their collaborational and computational aspirations.

References

1. Carlisle Adams and Steve Lloyd: Understanding PKI: Concepts, Standards, and Deployment Considerations, Addison-Wesley Longman Publishing Co., Inc., 2002
2. Enabling Grids for e-Science, http://www.eu-egee.org
3. Interactive European Grid, http://www.interactive-grid.eu/
4. Grid-Ireland, http://www.grid.ie
5. gLite, http://www.glite.org/
6. Burke, S., et al., The gLite 3.1 User Guide, April 2008, https://edms.cern.ch/document/722398/
7. Foster, I., Kesselman, C., Tuecke, S., 2001, The Anatomy of the Grid - Enabling Scalable Virtual Organizations, Int. J. High Perform. Comput. Appl., 15, 3, Sage Publications, Inc., pp. 200–222
8. Foster, I., Kesselman, C., 2003, The Grid 2 - Blueprint for a New Computing Infrastructure, Morgan Kaufmann Publishers Inc.
9. e-Infrastructure Reflection Group white paper, http://www.e-irg.org/publ/2006-Austrian-eIRG-whitepaper.pdf
10. The Apache Software Foundation, http://www.apache.org/
11. Globus, http://www.globus.org/
12. The Open Science Grid, http://www.opensciencegrid.org/
13. TeraGrid, http://www.teragrid.org/
14. NAREGI, http://www.naregi.org/

15. National Grid Service, http://www.grid-support.ac.uk/
16. Distributed European Infrastructure for Super Computers, http://www.deisa.eu/
17. NorduGrid, http://www.nordugrid.org/
18. Open Grid Forum, http://www.ogf.org/
19. Grid Interoperability Now, https://forge.gridforum.org/sf/projects/gin
20. Overview of the Grid Security Infrastructure, http://www.globus.org/security/overview.html
21. Witzig. C. SLCS and Vash, NREN and Grids Workshop,http://www.terena.org/activities/nrens-n-grids/workshop-06/slides/witzig-switch-slcs-vash.pdf, Malaga, Nov 07
22. The DataGrid Project, http://eu-datagrid.web.cern.ch/
23. The Large Hadron Collider Computing Project, http://lcg.web.cern.ch/LCG/
24. The LHC Experiments, http://public.web.cern.ch/Public/en/LHC/LHC-en.html
25. Andronico, G., Barbera, R., Falzone, A., Lo Re, G., Pulvirenti, A., Rodolico, A., 2003, The GENIUS web portal: grid computing made easy, ITCC '03: Proceedings of the International Conference on Information Technology: Computers and Communications, IEEE Computer Society, pp. 425–431
26. EnginFrame Grid Portal, http://www.enginframe.com/
27. Migrating Desktop, http://desktop.psnc.pl/
28. Sipos, G. Kacsuk, P. 2006, Multi-Grid, Multi-User Workflows in the P-GRADE Portal. J. Grid Comput., 3, 3–4, Kluwer Academic Publishers, pp. 221–238
29. GridSphere Portal Framework, http://www.gridsphere.org/
30. The Gilda Testbed, https://gilda.ct.infn.it/
31. OpenPBS Batch System, http://www.openpbs.org
32. Torque and Maui Batch System, http://www.clusterresources.com/
33. LSF Batch System, http://www.platform.com/
34. The Condor Project, http://www.cs.wisc.edu/condor/
35. Sun Grid Engine, http://gridengine.sunsource.net/
36. OGF Glue Schema version 1.3, http://forge.gridforum.org/sf/go/doc14185
37. MyProxy Credential Management Service, http://grid.ncsa.uiuc.edu/myproxy/
38. Carter, G., LDAP System Administration, O'Reilly Media Inc., 2003
39. Cooke, A.W., et al. The Relational Grid Monitoring Architecture: Mediating Information about the Grid, J. Grid Comput., Vol 2 No. 4, Kluwer Academic Publishers, pp. 323–339
40. VOMS Core User and Reference Guide, https://edms.cern.ch/file/571991/1/voms-guide.pdf
41. Workload Management System User and Reference Guide, https://edms.cern.ch/file/572489/1/EGEE-JRA1-TEC-572489-WMS-guide-v0-2.pdf
42. Logging and Bookkeeping User and Reference Guide,https://edms.cern.ch/file/571273/1/LB-guide.pdf
43. LCG File Catalog, https://twiki.cern.ch/twiki/bin/view/LCG/LfcGeneralDescription
44. FTS Users Guide, https://edms.cern.ch/file/591792/1/EGEE-TECH-591792-Transfer-Java-v1.0.pdf
45. AMGA Homepage, http://amga.web.cern.ch/amga/
46. Koblitz, B., Santos. N., The AMGA User's and Administrators manual, http://project-arda-dev.web.cern.ch/project-arda-dev/metadata/downloads/amga-manual_1_2_3.pdf
47. Global Grid User Support, http://www.ggus.org/
48. Grid Operations Centre DataBase, https://goc.gridops.org/
49. Service Availability Monitoring, https://lcg-sam.cern.ch:8443/sam/sam.py
50. Core Infrastructure Service, http://cic.gridops.org
51. The EGEE Training Digital Library, http://egee.lib.ed.ac.uk:8080/
52. The EGEE NA3 Homepage, http://www.egee.nesc.ac.uk/
53. ELGrid – an adaptive e-Learning tool for Grid education, http://www.grid.ie/elgrid/
54. JDL Attribute Specification (submission via WMS Network Server), https://edms.cern.ch/file/555796/1/EGEE-JRA1-TEC-555796-JDL-Attributes-v0-8.pdf

55. JDL Attributes Specification (submission via WMS WMProxy), https://edms.cern. ch/file/571273/1/LB-guide.pdf
56. The Storage Resource Manager, http://sdm.lbl.gov/srm-wg/
57. GSIFTP Tools for the Data Grid , http://www.globus.org/toolkit/docs/2.4/datagrid/deliverables/ gsiftp-tools.html
58. Childs, S. The MPI Working Group Report, http://egee-intranet.web.cern.ch/egee-intranet/NA1/TCG/wgs/EGEE-II-MPI-WG-TEC-2.pdf
59. GridSite, http://www.gridsite.org/
60. MySQL, Using Secure Connections, http://dev.mysql.com/doc/refman/6.0/en/secure-connections.html
61. GSI-Enabled OpenSSH, http://grid.ncsa.uiuc.edu/ssh/

AstroGrid and the Virtual Observatory

Nicholas A. Walton and Eduardo Gonzalez-Solares

1 Introduction

This chapter gives a brief introduction to the Virtual Observatory (VO). It looks in some detail at the UK's AstroGrid Virtual Observatory implementation, showing, via a range of science examples, how the "Virtual Observatory" can be of use to astronomers for a wide variety of data discovery and analysis tasks.

In Sect. 2 we describe a range of science and technological drivers which shaped the requirements in the development of the VO. The need for a set of standards to ensure that all VOs can fully inter-operate and how these were developed is discussed in Sect. 3. In Sect. 4 we begin the description of the AstroGrid project in more detail. Section 5 describes what the AstroGrid software infrastructure is, both from the end users point of view and from the angle of those wishing to use the software to publish their data and application resources to the VO. In Sect. 6 the actual use of the AstroGrid system by an end user astronomer is introduced. The topic of work-flows and scripting in the VO is described in Sect. 7. A range of worked examples of how AstroGrid and the VO can be used both to publish data and carry out science analysis is presented in Sect. 8, with Sect. 10 closing this chapter.

2 The Challenge of Data

The Virtual Observatory (VO) initiative is meeting the challenges resulting from the large new influxes of data in astronomy. Astronomy has historically been an observationally based science, where the study of the cosmos has allowed us to better understand the physical processes at work and allowed astronomers to answer a

N.A. Walton (✉)
Institute of Astronomy, University of Cambridge, Madingley Road, Cambridge, CB3 0HA, UK,
naw@ast.cam.ac.uk

E. Gonzalez-Solares
Institute of Astronomy, University of Cambridge, Madingley Road, Cambridge, CB3 0HA, UK,
eglez@ast.cam.ac.uk

Walton, N.A., Gonzalez-Solares, E.: *AstroGrid and the Virtual Observatory*. Lect. Notes Phys. **791**, 81–113 (2009)
DOI 10.1007/978-3-642-03370-4_3 © Springer-Verlag Berlin Heidelberg 2009

range of key questions such as how did the Universe form, how are galaxies created, under what conditions can planets form and so forth.

A range of powerful observatories have been constructed which are producing observational data across the wavelength domain. In recent decades, technological advances in areas such as detectors, have enabled the sky to be observed across the full range of the electromagnetic spectrum.

These new observational facilities, such as the Hubble Space Telescope[1] (HST) and the European Southern Observatory's (ESO) Very Large Telescopes[2] (VLT), bring with them significant data challenges. Data from these facilities come in many formats, with varying levels of complexities. The data is held globally at a range of major data archives such as the NASA Infrared Processing and Analysis Center[3] (IPAC), the European Space Agency's (ESA) science archives at the European Space Astronomy Centre[4] (ESAC), the European Southern Observatory (ESO) Science Archive,[5] and the Cambridge Astronomical Data Centre.[6] To add to the complexity, data is also to be found at a range of smaller archives and may also be published on "personal" web pages.

The trend in recent years, has been for researchers working on the data to be grouped into teams, formed to address specific research topics. The research collaborations can be of significant size (e.g. the Sloan collaboration [47]) and often contain researchers from a range of institutes, which in turn are located in differing countries.

A key challenge for the VO is in connecting these distributed research teams with the distributed and heterogeneous data and applications which they require in order to solve their particular science questions.

2.1 The Science Need for The Virtual Observatory

A number of major drivers have indicated the need for a "VO". These drivers are both scientific and technical. With the opening up of more of the electromagnetic spectrum to astronomers, the scientific questions that can be asked have become more complex.

For instance, observations in different wavelength regions can help address an understanding of differing processes. Take the well-known supernova remnant (SNR), Cassiopeia-A. Observations in the near infra-red reveal the distribution of dust in its nebula (e.g. recent Spitzer IRAC imagery [15]), whilst deep X-ray observations from Chandra enable the study of the shocks in the expanding nebula

---

[1] http://hubble.nasa.gov

[2] http://www.eso.org

[3] http://www.ipac.caltech.edu

[4] http://www.esa.int/esaMI/ESAC/index.html

[5] http://www.eso.org/sci/archive/

[6] http://casu.ast.cam.ac.uk/casuadc

(e.g. from deep multi-epoch observations [12]). Optical observations reveal the distribution of heavy elements in the SNR (e.g. from HST imagery [18]), whilst radio observations allow the mapping of the magnetic fields in the SNR (e.g. relationship of compact radio emission features with magnetic fields [3]). Merging the information from all data sources can enable a deeper understanding of the SNR as a whole.

The science roadmaps from the research communities in Europe are summarised on the AstroNet website [5], many of the key science topics requiring seamless access to multi-wavelength data in order to ensure significant progress. Indeed the Astronet Science Vision [6] notes that:

Astronomical data management and the Virtual Observatory:
With the ongoing expansion of instrumental capabilities across the electromagnetic spectrum, the importance of archival datasets and the combination of information from multiple datasets is growing rapidly. This trend will accelerate dramatically in the future, with the commissioning of dedicated survey facilities. The scientific output of these facilities will be maximized by making adequate provision for systematic archiving of the data, with common standards that allow for inter-accessibility of datasets and effective scientific exploitation of the data. The first steps in this direction are being made by individual data centres and through the European Virtual Observatory, which in turn is part of the International Virtual Observatory effort.[7]

Science drivers for the VO have been derived and published by a number of the VO projects (e.g. AstroGrid's initial development case [4] and the Euro-VO VOTECH project [42]), these reflecting major multi-wavelength science use cases.

2.2 The Technical Need for The Virtual Observatory

The range of new facilities in observational astronomy available since the 1960s is impressive – see the Wikipedia List of Observatories [45] which gives pointers to many of these facilities. The 1990s and into the 2000s have seen the opening up, at high resolution, of much of the electromagnetic spectrum. For instance Chandra and XMM-Newton in the X-ray, Spitzer in the Infrared, SWIFT studying gamma ray bursts, SOHO studying our sun at a range of wavelengths, to name but a few. Large ground-based observatories such as the ESO VLT, where the four 8.2-m diameter telescopes are each equipped with three major (and different) instruments, generate very large and complex data flows on a nightly basis.

These observational facilities present a range of challenges. There is more data [30] (at the peta scale), more complex data (objects are routinely described by hundreds of parameters) and data from more multiple sources. The data itself is diverse: images, spectra, three or more dimensional data, tabular data and so forth. Data is described in a range of manners, with differing units and errors.

Astronomers are realising that increasingly sophisticated analysis of the full range of data at hand is required to allow discrimination between competing theoretical models. Further, new facilities being planned are required to be global in

---

[7] Published with permission.

scale, with a corresponding requirement on the community to plan to exploit the data from these facilities to the fullest extent, through making the data available to the widest possible community, in an accessible format, as quickly as possible.

No longer is it acceptable for a research astronomer to have to become an expert in the reduction and acquisition of data from the many wavelength realms in astronomy. More time must be made available to the astronomer to interpret the observational data to enable the continuing output of scientific understanding.

Thus, the Virtual Observatory initiatives commenced towards the end of the 1990s with the aim to meet these science and technical challenges.

3 Standards and Interoperability

In order to create an operational data infrastructure, the VO projects in a number of countries, such as the National Virtual Observatory (NVO) in the USA [26] and AstroGrid in the UK, realised at an early stage the importance of interoperability standards. The International Virtual Observatory Alliance [21] (IVOA)[8] was formed in June 2002 where the necessary technical standards are formulated and agreed upon. The VO projects, implementing their systems conforming to these standards, then ensure full interoperability. In short this means that data published through VO compliant standards in say the USA will also be fully visible and usable by a user accessing those data through a VO system in Europe, China, etc.

The IVOA work is carried out in a number of working groups. These cover areas such as

- *Standard vocabulary:* standards in this area are aimed at, for instance, creating a dictionary for astronomy and a list of standard terms to describe astronomy concepts (so-called UCDs, Unified Content Descriptors).
- *Standard ontology:* describing how terms are related.
- *Standard data models:* (encoding format) for each type of measurement. VOTable[9] for instance is an XML format defined to allow for the interchange of tabular data in the VO.
- *Standard query language:* for issuing spatial, temporal and semantic queries across the catalogs. The main current (2008) effort in this area is in developing the Astronomical Data Query Language.[10]
- *Standard access services:* for retrieving catalog records or image cutouts. Standards in this area include the Simple Image Access Protocol (SIAP)[11] which allows access to image collections.

---

[8] http://www.ivoa.net

[9] http://www.ivoa.net/Documents/latest/VOT.html

[10] http://www.ivoa.net/Documents/latest/ADQL.html

[11] http://www.ivoa.net/Documents/latest/SIA.html

- *Standard mechanisms for interacting with storage systems:* this is related to access and availability of distributed storage systems, VOSpace[12] having been developed to allow for "staged" data.
- *Standard authentication/authorisation mechanisms:* essential to allow for group access to data that may be proprietary.
- *Standard event notification services:* allowing for the rapid publication of science alerts to support for instance alerts describing new gamma ray burst discoveries, which in turn require rapid follow-up. The standard in this area in VOEvent.[13]

The IVOA operates a vigorous standards process [22] modelled on that adopted by the W3C foundation [46]. Upon adoption, the IVOA recommendations are published; these are then implemented by the VO projects. The list of documents published by the IVOA is available at http://www.ivoa.net/Documents/.

4 AstroGrid: The UK's Virtual Observatory

As noted previously in Sect. 3, AstroGrid[14] [41] was one of the founding members of the IVOA, starting at the end of 2001. It is a consortium of leading UK astronomy institutes with a background in astronomical data processing or curation. As of 2008 the consortium consisted of The University of Bristol, The University of Cambridge, The University of Central Lancashire, The University of Edinburgh, The University of Leicester, The University of Manchester and the Rutherford Appleton Laboratory. The key aims of the AstroGrid project were encapsulated in its mission aims:

- Improve the quality, ease, speed and cost effectiveness of on-line astronomy
- Make comparison and integration of data seamless
- Removing barriers to multi-wavelength astronomy
- Enable access to very large data sets

The development phase of the project ran from 2001 to 2007, with a deployment period for the UK running into 2009. AstroGrid made a number of development releases of software in 2006 and 2007. However, the first full v1.0 release occurred on 1 Apr 2008 and it is this software which is described more fully in the following sections.

4.1 AstroGrid and the European Virtual Observatory

AstroGrid is one of the key partners in the European Virtual Observatory (Euro-VO). The Euro-VO is organized into three main elements:

---

[12] http://www.ivoa.net/Documents/cover/VOSpace-20080124.html

[13] http://www.ivoa.net/Documents/latest/VOEvent.html

[14] http://www.astrogrid.org

- *Euro VO Technology Centre (VOTC)*: this activity is led by AstroGrid and is charged with developing the technical infrastructure of the Euro-VO. Currently (2008) the VOTC is funded through the EU FP7 VOTECH[15] project [43].
- *Euro VO Facility Centre (VOFC)*: this activity is jointly lead by ESO and ESA and operates the Euro-VO service and training activities.
- *Euro VO Data Centre Alliance (DCA)*: this activity, led by CDS Strasbourg, organises the main data centres in Europe and coordinates their activities in deploying VO interfaces. Currently (2008) the DCA is funded by the EU FP6 Euro-VO DCA[16] project.

The Euro-VO AIDA[17] EU FP7 project is, from 2008, providing a limited amount of resourcing for all Euro-VO activities. Full details can be found on the Euro-VO website.[18]

In the context of the following subsections we refer to AstroGrid infrastructure components. However these components also form the basis for the wider Euro-VO infrastructure (with additional components integrated from partner Euro-VO projects) which will be formally released through the VOTECH project at the end of 2008. Thus, in the future, regard AstroGrid and Euro-VO as interchangeable.

5 AstroGrid: The Infrastructure

The AstroGrid software is Java based and is structured around a set of intercommunicating web services. It has an asynchronous architecture which is necessary for running work-flows using the distributed services available through the VO. Astro-Grid has developed its system to give the end user single-point access to globally held data sets, recognising that the availability of applications to seamlessly work with those data is also important.

The AstroGrid software implements the various IVOA interoperability standards. Further it exploits the wider recent developments in distributed computing and makes use of the availability of high-speed networks linking the various astronomical data centres.

AstroGrid has two aspects: an infrastructure which runs server side to allow for the publishing and manipulation of data and then a small client side application which an end user (astronomer) can use to access these VO services.

The key server-side infrastructure components are as follows:

- *Registry:* A database listing meta-data about available services, including worldwide data services, applications that can be run and internalAstroGrid services.

---

[15] http://www.eurovotech.org

[16] http://cds.u-strasbg.fr/twikiDCA/bin/view/EuroVODCA/WebHome

[17] http://cds.u-strasbg.fr/twikiAIDA/bin/view/EuroVOAIDA/WebHome

[18] http://www.euro-vo.org

IVOA compliant registries around the world can harvest from each other. The registry concept is a fundamental underpinning of the VO; once a resource is "published" in the registry, it is then possible to access that resource globally. The IVOA Registry of Registries[19] gives the locations of these inter-operating registries.

- *VOSpace:* Virtual storage. Sites offering bulk storage run a "VOSpace Server", whilst those sites also offering access run a "VOSpace Manager". VOspace offers the ability for research teams to share data and work-flows with colleagues and in the longer term will offer individuals or groups the possibility to publish their own data and results.
- *Data Set Access (DSA):* This is a software intended to enable data providers methods to publish data through standardised VO interfaces.
- *Common Execution Architecture (CEA):* This is the framework which allows for the execution of applications within the VO. Each application is registered in the "Registry" and thus is discoverable by any user of the VO.
- *Astro-Taverna:* In order to provide a graphical work-flow system the Taverna work-flow system [37] has been used as a base infrastructure, thus providing a client side wok-flow editor and engine. Taverna can invoke various flavours of service – ranging from local java classes, through standard WSDL-described web services, to various "grid" services. It was developed to meet the needs of the bioinformatics community and thus has access to many bioinformatic services. In order to utilise Taverna with AstroGrid, a processor plugin [44] (Astro-Taverna) has been developed which exposes all the functions of the AstroGrid AstroRuntime (AR) as services in Taverna.
- *Community:* This is a software for administering users and transmitting identification information to allow single sign on. It is installed and run at local sites. Users can then, for example, define a group of collaborators who can utlise that sites VOspace installation.

From the client side the following components are available:

- *VODesktop:* VODesktop is the launching point for doing science using VO tools. It uses a resource centred approach. First you choose the resources you are interested in by searching the registry of resources. Next, you query the chosen resources to look for and fetch the specific data you want. Finally, you either save the data (to VOSpace or to local disk) or pass the data on to another application (e.g. an image viewer) to start doing some science analysis.
- *Astro Runtime:* This is effectively the AstoGrid VO middleware. It provides a set of standard interfaces to the various VO services. These API's can be programmatically accessed from user side applications through SOAP[20] or HTTP. This simplifies VO programming as developers gain easier access to all VO

---

[19] http://rofr.ncsa.uiuc.edu/cgi-bin/rofrhello.py
[20] SOAP: a protocol for exchanging XML based messages.

services, through the astro-runtime interface. VODesktop itself is built on the Astro-Runtime.[21]

The following subsections describe the AstroGrid server side software in more detail, whilst the user side client software is described in Sect. 6. We describe in more detail the technologies involved in publishing data and then proceed with a brief overview of the other VO components.

5.1 Publishing Data

In collaboration with others in the International Virtual Observatory Alliance (IVOA), AstroGrid has participated in developing and implementing a set of standards defining:

- a common data description Language, VOTable;
- a common data set query language, ADQL;
- a set of standard Web Service interfaces for accessing data sets.

The AstroGrid Data Set Access Component (DSA) runs in a separate Web Service container, typically Apache Tomcat,[22] and does not require changes to the underlying database system. This means that data centres can use the AstroGrid DSA to publish their data to the VO without having to modify their existing database systems.

DSA exposes selected tables in the database. It does not provide standardised views of the database and it does not require the publisher to change the database in any way. It needs only a read-only connection to the database, so there is no risk of damage to the data.

On the input side, the AstroGrid DSA provides a translation layer between the platform-independent ADQL (see Sect. 3 and the specific flavour of Structured Query Language (SQL) required by the database which hosts/manages the data resource).

ADQL provides a uniform query language for all data sets within the VO, which insulates the astronomer from having to understand the particular flavour of SQL required by each individual data set. This enables astronomers to create generic ADQL queries and send the same query to multiple data sets on different platforms and database systems.

On the output side, the AstroGrid DSA provides a translation layer to generate a VOTable representation of the results from the target database system. This provides a standard data format enabling the results of a database query to be used in other VO components. Results are stored in the AstroGrid VOSpace system, making them available to subsequent steps in the work flow.

---

[21] http://www2.astrogrid.org/desktop/astro-runtime

[22] http://tomcat.apache.org/

DSA has two web service interfaces which allow for "simple" and "complex" queries of the data to be undertaken. The "simple" query conforms to the IVOA "cone search" and supports simple positional queries of the database, where all results for objects specified by the the positional query will be returned. The second interface supports full ADQL queries thus allowing sophisticated and fine-grained queries of the database – important for cases where the data sets are large.

The Web Service interface provided by the AstroGrid DSA allows data set queries to be included as steps in a complex work-flow, sending the results from a data set query into the inputs of another step. As an example, astronomers can create a work-flow which used a CEA-wrapped instance of an application which identifies sources in a set of images, query a number of different data sets for known sources within the field of the images and then combine the query results with the extracted sources from the images in another work-flow step to re-register the images based on the positions and proper motions of the known sources.

Astronomers can use the same query builder tool in AstroGrid to build ADQL queries for each of the data sets involved, without having to know anything about the underlying database platforms at each remote service location.

This kind of activity represents a major step towards one of the overall goals of the VO of providing astronomers with access to data from a wide range of data sources distributed across the global grid of data centres and data sources.

Other forms of data require additional data publishing software. Much of these have been developed by Euro-VO partners and are noted in Sect. 9.2.

5.2 Publishing Applications

The AstroGrid CEA [20] is a way to provide science applications, either legacy code or new, as services. Applications can be wrapped, thus configured to run on demand in an application server. Wrapped applications can be called from desktop applications, work-flows or standard VO services. They are registered in a IVOA standards compliant registry; hence, they can be discovered and used dynamically by CEA-aware user interfaces.

The CEA package provides

- application-servers for various kinds of applications;
- server class-library and framework for writing new kinds of application servers;
- a client class-library for calling CEA services.

CEA applications interface to VOSpace servers; thus applications can store their outputs to local disk storage, facilitating staging of intermediate data products. The CEA provides methods to delegate processing to grids and thus supports computationally intensive use cases.

5.3 Managing Communities and Security/Authorisation

Much data accessible through the VO is public in nature. However, to support collaborative work and make access to authorised users possible in the case where the data is still, for example, in a proprietary period, AstroGrid has developed a community/authorisation infrastructure [33]. The community service allows for single sign on; subsequent access to all authorized resources then occurs without the need for further password checks.

Authentication of agents to services uses X.509 identity certificates in the end-user's name. SOAP requests are secured at message level using digital signatures according to the OASIS WS-Security standard. Other protocols (e.g. ad-hoc application protocols carried over HTTP) use the client-authentication feature of standard Transport-Layer Security.

Services accept end-user certificates issued by trusted Certificate Authorities (CAs); the users do not need to preregister their public keys at each service. IVOA mandates a global Single Sign-On (SSO) system for the International Virtual Observatory (IVO), based on interoperable Public Key Infrastructures (PKIs) in the regional virtual-observatory projects. AstroGrid counts as an IVO region. Service providers in AstroGrid are asked to accept certificates from equivalent CAs in other regions. Currently, the CA arrangements are not finalised, but AstroGrid will probably use the UK e-Science CA.

End-users are registered into the AstroGrid PKI at on-line astronomical communities where a community typically represents a single astronomy department. These communities are the registration authorities (RAs) for the national CA. The communities also operate services that provide short-term ('proxy') certificates to user agents following the MyProxy protocol. It is intended that end-user's long-term certificates and keys are held by the communities and not managed by the users themselves.

Authorisation in VO services is not currently standardised by the IVOA. Service providers define and manage their own authorisation tables and have complete control over authorisation decisions. Most access rights are assigned to user groups. AstroGrid supports and prefers groups defined inside communities by the RAs.

5.4 The Registry

VO resources are discovered through a registry service which provides considerable meta-data (thus descriptions) for each resource. For example, the schema for registering databases allows the full Data Base schema to be described in detail, which means that when the user accesses that database they have full information about the structure of that resource. The AstroGrid registry fully conforms to IVOA standards.[23]

---

[23] http://www.ivoa.net/Documents/latest/RegistryInterface.html

Data publishers work with the AstroGrid publishing registry in that they can add the meta-data for their resources to the collection and make those resources known (publish) to the VO. The AstroGrid registry is a full "harvesting registry" which collates information from all known publishing VO registries, for instance in USA, China, etc. Thus, end-users working with the AstroGrid system have access to the full range of global VO data and application resources.

5.5 VOSpace: Data Staging, Distributed Data Storage

VOSpace is the distributed storage system developed for use with the Virtual Observatory. It is used with VO services such as DSA and CEA to allow for data staging. This is important in that it allows data to be exchanged between services in work-flows without copying those data to and from the user's desktop. For example, the results of queries on two, perhaps very large data databases, might be written to files in VOSpace and then read into a cross-matching service which then writes its results to a subsequent file which is then available as input for a following application. VOSpace thus facilitates the concept of server side science work-flows, where data, especially intermediate data, is moved around on the fast backbone networks, with only the end results actually being drawn down by a the user to their local client machines.

The AstroGrid VOSpace implementation is fully compliant with the IVOA standard.[24] VOSpace is also an interface to data grids. For instance the iRODS [31] storage resource system[25] now has a VOSpace interface, meaning that users can use this to manage potentially extremely large data collections and pass those data to data grids which already interface to iRODS installations.

6 VODesktop: The Interface to the Virtual Observatory for the Astronomer

This section gives details on the client side software which AstroGrid provides for the astronomer to allow access to VO data and applications. The key user software application is the VODesktop graphical user interface which provides a range of functionality, allowing one to fully access and exploit VO data services. It is worth noting here that the resources available in the VO are typically web services rather than web pages, thus are not designed to be directly accessible via a web browser. Hence the use of a tool, such a VODesktop, provides the user interface to these services.

---

[24] http://www.ivoa.net/Documents/cover/VOSpace-20080311.html
[25] http://www.irods.org

6.1 Finding Resources in the VO: VOExplorer

The VOExplorer application of VODesktop is used to search for resources in the Virtual Observatory; it does this by querying the AstroGrid Registry which is a service which stores (and regularly updates) the details of all of these resources.

Because there are now thousands of resources available through the VO, the VOExplorer tool gives the user a number of options to carry out their searches. The VOexplorer tool has a number of panes – see Fig. 1.

1. The Smart and Static Lists pane (top left) displays the names of resource lists which have been used to query the AstroGrid registry.
2. The Resource Lists pane (top right) lists the resources.
3. The Information, Table Metadata and XML pane (bottom right) displays information on any resource selected in the Resource Lists pane.
4. The Actions and About pane (bottom left) lists the actions that can be invoked on the selected resources (clicking an action invokes it).

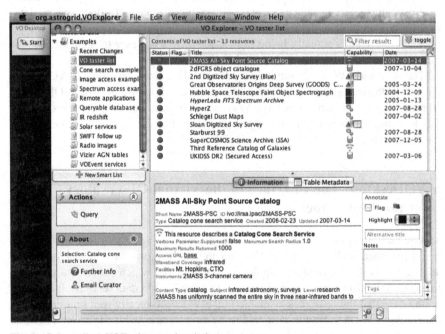

Fig. 1 (Color online) VOExplorer main window

A typical session with VOExplorer might involve the four panes in the order listed: create a list of resources using the Smart and Static Lists pane, select a resource in the Resource Lists pane, read a description of the resource in the Information, Table Metadata and XML pane and invoke an action on the resource using the Actions and About pane (Fig. 2).

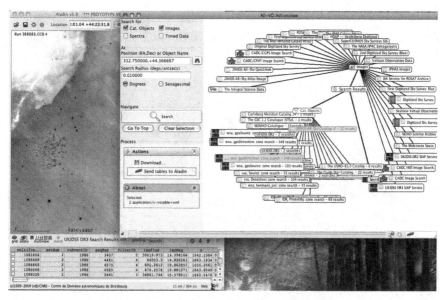

Fig. 2 (Color online) AstroGrid VODesktop in use. This shows the result of a full AstroScope query around the position of the nebula IC 5070. Image data from the IPHAS survey are displayed in the Aladin client, either corresponding catalogues over plotted

Once a resource has been selected the user is offered a range of actions that they can perform upon that resource. The range of actions presented is automatically determined from the type of service/resource. Actions include:

- Accessing Databases via the built in *Query Builder* interface. This allows for SQL queries to be constructed and sent to the database, with the results of that query being sent to the users VOSpace. An example of this in use is shown in Sect. 8.2.2.
- Running Applications via the VODesktop built in *Task Runner* interface. If the user selects an application, then they are able to execute that application remotely, with results output to the users VOSpace.

6.2 Finding all Data: AstroScope

A common question is, what image data, catalogue data, etc., is there available at a particular location on the sky. The Astroscope application (in All-VOScope mode) allows for a user defined search box to be defined, with queries then being automatically sent to all VO services, with the list of those resources with catalogues, spectra, images and time series data available at that position returned to the user. They are then able to select the data of interest and carry out further analysis of those data. All-VOScope significantly speeds the ability of the astronomer to discover data, as

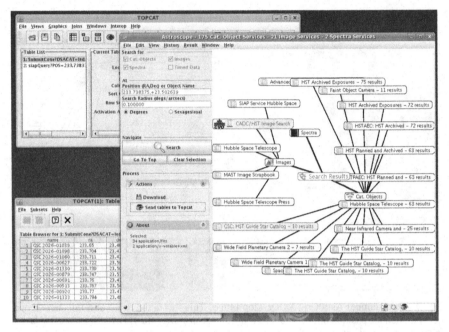

Fig. 3 This image shows All VOScope in use, where a range of data resources are visualised. In this example a selection of catalogues have been selected, with some of these loaded into the TopCat tool

it avoids the user having to interrogate many different back-end archives on a one by one basis (Fig. 3).

6.3 VODesktop: Other Applications

VODesktop also has a range of "housekeeping" applications. These include the "File Explorer" which allows the user to navigate their VOSpace storage area. Also, if the user selects a data file, they are able to simply transfer that file into a relevant client side visualiser via "PLASTIC" tools such as Aladin [10][26] (see discussion in Sect. 6.4).

Further tools include a self-test function which checks network connectivity, runs memory tests and so forth.

6.4 Desktop Client Interoperability: Plastic

VODesktop supports the PLASTIC[27] [38] messaging system that provides inter-operability between applications running on the desktop. All the PLASTIC-aware

---

[26] http://aladin.u-strasbg.fr/

[27] plastic.sourceforge.net

applications on a desktop can share data. Thus for example data can be loaded into a tool that visualises tabular data, a sample can be selected there and sent to another tool directly, perhaps overlaying selected catalogue data points over a corresponding image. Over the course of 2008–2009, the PLASTIC protocol is being generalised into a full IVA complaint standard named Simple Applications Messaging Protocol (SAMP).[28]

7 Workflows in the VO

This section describes how work-flows can be created and executed utilising the AstroGrid system. A work-flow is simply a sequence of VO-related tasks, possibly including loops, which can be saved in some format and then re-run. An example might be querying a specific "SIAP" image service to find an image at a particular RA and Dec; sending the image found to an invokeable copy of an image source extractor (such as SExtractor [9]), putting in suitable parameters and deriving an object catalogue; saving the resulting table to your VOSpace; and then also loading the catalog into a desktop client able to visualise tabular data such as TopCat [39]. Before re-running the work-flow, one can then change the parameters used or otherwise alter some of the tasks; for example running against a different SIAP service or using a different RA and Dec. There are two ways of building a work-flow using AstroGrid tools.

7.1 Astro-Taverna: Graphical Work-Flows

The first way is with an application that uses a graphical interface that puts together standard building blocks visually. The work-flow is saved as a work-flow document in a standard XML format; the application can also load such a work-flow document and alter the visual flow diagram. This approach has become common in the world of Bioinformatics, from which the known tool "Taverna" [37] was developed. AstroGrid has been developing a version of Taverna with VO plugins, named "Astro-Taverna" [44]. This will be made available in the next AstroGrid release towards the end of 2008. Figure 4 show an example of Astro-Taverna in use, automating the query of two databases and cross matching the outputs.

7.2 Scripting AstroGrid with Python

In this section we give some simple examples of how the user can interact with the Virtual Observatory through the AstroGrid Python scripting interface, the "second" way to automate sequences of VO tasks being to write a script that includes calls that correspond to VO tasks. An example (using AstroGrid Python) is

---

[28] http://www.ivoa.net/Documents/latest/SAMP.html

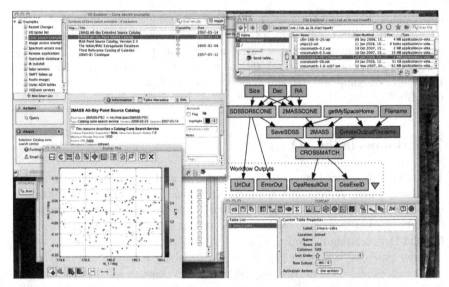

Fig. 4 (Color online) This shows the result of running a simple Astro-Taverna work-flow where two databases are queried and the outputs of these queries then cross-matched. The final results file is stored in the users VOspace storage area and is shown here displayed in a client visualiser

```
>>> img = m.readfile('#sdss/image.fits')
```

which reads a file called `image.fits` from a folder in the user's VOSpace called `sdss` and puts it into an object called `img`. In principle one could invent a complete new "VO command language". However, our approach is rather to define a series of function calls within the context of an existing language, so that the user can embed VO tasks within the context of more general programming and processing. Python[29] is an ideal choice, because of its growing popularity, the existence of other scientific and mathematical modules and because it is built to make accessing the internet simple.

7.2.1 GUI vs CLI

When VODesktop is launched it starts a core service called AstroGrid Runtime (AR). The AR serves as an interface between the user and the VO, receiving the requests from VODesktop and sending back the results. The AR has been designed to be independent of VODesktop in the sense that any other tool can connect to it in the same way that VODesktop does making possible then to write scripts which perform the same actions or even extend the possibilities of VODesktop. In this section we will describe how to access the AR from the command line interface

---

[29] http://www.pyhon.org

(CLI) in particular using Python. There are several benefits from using the command line to access the VO. As mentioned above you can effectively write work-flows or series of VO calls to do repetitive tasks. By having all your commands written in a file you have a transcript of the steps you have followed to reach to a particular result and you can repeat the process. Also you can embed calls to VO services from your own programs so that the VO access is just another call in a more complex task integrating the VO into the user scripts.

7.2.2 Starting Python

In order to make the instructions simpler we assume that the user is running a Unix-like operative system and has access to a command line running a bash-like shell. Instructions for other operative system or shells will vary slightly and are included in the documentation of the packages described.

The first step is to start the Python interpreter. In order to do this one generally types python in the shell command line. This is the standard interpreter bundled with every Python installation but lacks some features of other interactive environments. For this purpose we recommend the use of IPython[30] which adds many more features for interactive analysis (TAB completion, command history, session logging, aliases, debugger, etc.). Once installed (see below for instructions) this is generally invoked by typing ipython in the command shell.

7.2.3 Installing Additional Packages

Standard Python installations lack some of the powerful modules needed in order to perform numeric analysis and data plotting. Although it is not necessary to install them to start using AstroGrid Python they will be useful later when analysing the results.

The first thing is to define where the additional modules will be installed. This will be done in a file ~/.pydistutils.cfg located in each user home directory. Its content is

```
[install]
install_lib = ~/Library/Python/2.5/site-packages
install_scripts = ~/bin
```

where install_lib is the location of the directory we will be installing the packages to (remember to create that directory beforehand), assuming that Python version 2.5 is installed. After this we will need to set the PYTHONPATH enviromental variable to this directory, e.g.

```
export PYTHONPATH=~/Library/Python/2.5/site-packages
```

---

[30] http://ipython.scipy.org/

It is useful to add this line to the user's shell initialization file (e.g. ~/.bashrc). Then we are using EasyInstall[31] to install the additional Python modules:

```
wget http://peak.telecommunity.com/dist/ez_setup.py
python ez_setup.py
```

This will create a script in the user's bin home directory which will make very easy to install additional modules. Once this is done, additional packages can be installed simply by typing, e.g.

```
~/bin/easy_install ipython
```

to install the IPython interpreter or

```
~/bin/easy_install numpy
```

in order to install the numeric Python extension called numpy.[32] Other useful modules for astronomy are PyFITS[33] to work with FITS files and matplotlib[34] to produce plots and interact with them. The full list of packages available for Python is available at http://pypi.python.org/pypi.

7.2.4 AstroGrid Python

From Python we can access the interface provided by the AR using the XMLRPC[35] protocol. In AstroGrid we have written a Python module which hides from the user the complexities of XMLRPC and performs the calls to the AR transparently. This is installed as

```
~/bin/easy_install astrogrid
```

Once installed start Python and VODesktop. The latter is needed because we are making calls from Python to the AR which is started when VODesktop is launched.

The following example shows a very simple set of commands which query the NASA Extragalactic Database (NED) for all known objects around a particular position. In Astronomy this kind of search is called a cone search.

In the examples below we will show all the typed lines starting with the standard prompt of the Python interpreter (>>>) and the continuation prompt (. . .).

```
>>> from astrogrid import ConeSearch
>>> ivorn = "ivo://ned.ipac/Basic_Data_Near_Position"
```

---

[31] http://peak.telecommunity.com/DevCenter/EasyInstall

[32] http://numpy.scipy.org/

[33] http://www.stsci.edu/resources/software_hardware/pyfits

[34] http://matplotlib.sourceforge.net/

[35] http://www.xmlrpc.com/

```
>>> cone = ConeSearch(ivorn)
>>> ra, dec, sr = 242.811, 54.596, 0.1
>>> result = cone.execute(ra, dec, sr,
...                        saveAs='#cones/ned.vot')
```

After starting the Python interpreter we need to load the libraries that we want to use. The first line of the example does precisely this; it imports the ConeSeach class from the AstroGrid library. There are different ways of importing libraries in Python:

```
>>> # Import the whole math library in the current namespace
>>> # Functions will be called e.g. radians()
>>> from math import *
>>> # Import only some constants and functions from the library
>>> from math import pi, radians, degrees
>>> # Import the math library
>>> # Functions will ba called e.g. math.radians()
>>> import math
>>> # Import the math library and rename the namespace
>>> # Functions will be called e.g. m.radians()
>>> import math as m
```

The second line of our example defines the variable `ivorn`. We are used to Uniform Resource Locators (URL) which identify the address of a web service (e.g. http://www.astrogrid.org). In the VO these are called IVORN (IVOA Resource Name) and serve the same purpose than the URLs, i.e., identify the location of a service. In short the NED service publishes in the web to http://nedwww.ipac.caltech.edu and in the VO their cone search service is published as ivo://ned.ipac/Basic_Data_Near_Position. Once defined the service we want to query the third line creates a cone search instance `cone` which points to the service. Once the R.A. and Dec coordinates of the centre of the cone and its radius are defined the method `cone.execute` is used to perform the cone search and save the results as a VOTable in the user's VOSpace. The table can be downloaded to local disk for inspection or visualised in TopCat. This can be easily accomplished by launching TopCat and typing

```
>>> from astrogrid import acr
>>> acr.startplastic()
>>> acr.plastic.broadcast(result)
```

in Python, where `result` is the output variable from the `cone.execute` method above.

Of course using the command line to access the VO provides a lot of flexibility and capabilities not found in VODesktop but this comes with a price: one has to read the manual and know the name of the commands. This is in some way alleviated by the interactive help available from within Python. In IPython you can type

```
>>> help(ConeSearch)
```

in order to get help on the ConeSearch class, usage examples and available method calls. In particular the output of the above command is

```
Help on class ConeSearch in module astrogrid.cone:

class ConeSearch
 |  The following example sends a cone search query to
 |  NED and saves the resulting VOTable in the local
 |  disk.
 |
 |    >>> from astrogrid import ConeSearch
 |    >>> iv = "ivo://ned.ipac/Basic_Data_Near_Position"
 |    >>> cone = ConeSearch(iv)
 |    >>> print cone.info['content']['description']
 |    >>> result = cone.execute(242.811, 54.596, 0.1)
 |    >>> open("ned.vot",'w').write(result)
 |
 |  :IVariables:
 |    info
 |       Information about the service
 |
 |  Methods defined here:
 |
 |  __init__(self, service)
 |      :Parameters:
 |        service : str
 |           URI of service to be queried
 |
 |  execute(self, ra, dec, radius, saveAs=None,
 |               clobber=False)
 |      Execute the cone search.
 |
 |      :Parameters:
 |        ra : float
 |           R.A. in degrees
 |        dec : float
 |           Dec in degrees
 |        radius : float
 |           Radius in degrees
 |
 |      :Keywords:
 |        saveAs : str
 |           Saves the query to a file in MySpace.
 |           Default: None
 |        clobber : bool
 |           Overwrites file if it exists
 |
 |      :Return:
 |        res : str
 |           VOTable as a string or the name of
 |           the output file if 'saveAs' was used.
```

This example is equivalent to selecting the NED resource in VODesktop and performing a query using AstroScope. The power of scripting comes to play however when the user wants to perform this task for a number of positions. One can easily think of extending this script to read a list of object coordinates from a file and perform one cone search for each object saving the result of each query to a different file.

This is illustrated in the following script which queries the Sloan Digital Sky Survey database for a list of 20 positions.

```python
#!/usr/bin/python
"""This example shows how to submit a list of
   ra,dec positions to the SDSS DR5 cone search.
"""

import sys, time
from random import random
from math import degrees, pi

from astrogrid import ConeSearch

# Define the service endpoint IVORN
ivorn='ivo://wfau.roe.ac.uk/sdssdr5-dsa/dsa'
# Create the cone search instance and define
# the table that we are going to query
# (this is only necessary in multitable databases)
cone=ConeSearch(ivorn, dsatab='PhotoObjAll')

# Generate 20 random positions.
# In real life these would be read from a file
nsrc=20
ra=[random()*pi*2*degrees(1) for i in range(nsrc)]
dec=[(random()*pi-pi/2.)*degrees(1) for i in range(nsrc)]
radius=[0.001]*nsrc

for i in range(nsrc):
        res = cone.execute(ra[i], dec[i], radius[i])
        open('sdss%02d.vot' % (i+1), 'w').write(res)
        print i+1, 'sdss%02d.vot' % (i+1)
```

In order to execute this script save it to a file, e.g. `conetest.py` and from the command shell execute

```
python conetest.py -b
```

The script will create a file in the current directory holding the output result of the query for each object.

7.2.5 Additional Information

We have described a simple case usage of AstroGrid Python but basically everything that can be done from the VODesktop user interface can also be scripted. This
includes searches to image archives and image download (included visualisation in,
e.g. Aladin), remote database queries using ADQL and running remote tasks (e.g.
extracting objects from images or catalogue cross match).

There are many resources in the Web for learning Python, including manuals
on using Python for astronomy. The AstroGrid web pages[36] also provide extensive
documentation on using the AstroGrid Python module as well as example scripts to
get the user started.

8 AstroGrid in Use

In this section we give a number of examples of how the AstroGrid software is
already in use, both for publishing data and supporting science analysis.

8.1 Data Publishing

In order to make a catalogue available to the VO it first needs to be stored in a
database system. At the moment the most popular database systems are supported
by our software (PostgreSQL [29], MySQL [25], Sybase [36] and Microsoft SQL
Server [24]). After this is done the AstroGrid Data Set Access (DSA) library provides the layer to publish the catalogue in a VO standard way.

Here we give two specific examples of how AstroGrid technology has been used
to publish data. These are the IPHAS [14] and Hipparcos [16, 28] data sets.

8.1.1 The IPHAS Catalogue

The INT Photometric $H\alpha$ Survey of the Northern Galactic Plane (IPHAS; [14]) is a
$1800\,deg^2$ CCD survey of the northern Milky Way ($|b| < 5°$) using the r, i and $H\alpha$
pass-bands down to $r = 20$ (Vega, 10σ for a point-like source in an aperture of 1.2
arcsec). The Initial Data Release (IDR; [19]) of the IPHAS survey contains observations for about 200 million objects and comprises 2.4 Tb of processed imaging
data. Figure 5 shows the coverage of the catalogue in galactic coordinates.

The data is processed at the Cambridge Astronomy Survey Unit (CASU) producing astrometric and photometric calibrated images and object catalogues in FITS
format. This FITS catalogues are converted to ASCII (comma separated values)
files for ingestion in a Sybase dataserver running in a single machine with four
Xeon processors and 4 GB of memory. The data ingestion is accomplished in a

---

[36] http://www.astrogrid.org

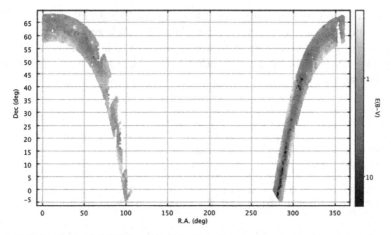

Fig. 5 (Color online) IPHAS IDR coverage in Galactic coordinates. The intensity of the colour reflects the Galactic extinction as given by the Schlegel maps [34]

two-step process. First the single band and band-merged catalogues are read and one ASCII file per database table in the database is written to disk. In a second stage the ASCII tables are ingested in the database and proper indexes created. There are two separate user interfaces to the IPHAS IDR. Users who only want to extract fluxes or images around a small position of sky (there is a limit on the radius of 1 degree) will find the web interface useful.[37] Given a position or an object name the user will obtain a cone search of the catalogue in a variety of output formats (VOTable, Binary VOTable, Comma Separated Value or HTML). Input lists of objects are also supported. Together with the IPHAS catalogue we provide access from the same interface to other useful VO compliant archives like SDSS DR5 [2], 2MASS Point Source Catalogue [35], NED and the Carlsberg Meridian Astrometric Catalogue [17]. Postage stamps generation for a particular input position is also available using the finding chart utility.

Users who want more sophisticated queries on flags and algebraic combinations of parameters (e.g. colours) will find the AstroGrid *Query Builder* more suited to their needs. The *Query Builder* is an application inside the AstroGrid VO Desktop which allows to submit an arbitrary query to the database system. The *Query Builder* facilitates the task of building complex queries written in Astronomical Dataset Query Language (ADQL [27]; an international agreed standard based on the Structured Query Language – SQL), save the queries and submit them asynchronously to the server (this facilitates sending long-running queries and retrieve the results later).

For example the next query returns objects classified as point-like in the r band and observed in conditions of seeing better than 1.5 arcsec. The main catalogue table is called *PhotoObjBest* and contains information about the detected objects

---

[37] Available from http://www.iphas.org/idr

(magnitudes, positions and other image derived parameters) while the table *Chip-sAll* contains the data quality information (like magnitude zero point, seeing and other observing conditions). We then have to join both tables in order to select the information needed. The table *tGetNearbyObj* is a helper table used to perform an exact cone search.

```
SELECT P.ra, P.dec,
   P.coremag_r, P.coremag_i, P.coremag_ha,
   P.coremagerr_r, P.coremagerr_i, P.coremagerr_ha
FROM PhotoObjBest as P, tGetNearbyObj as G,
   ChipsAll as CR
WHERE G.ra=300.0 AND G.dec=30.0 AND G.r=10.0
   AND CR.seeing<1.5 AND (P.class_r=-1 OR P.class_r=-2)
   AND P.objID = G.objID AND P.chips_r_id=CR.chipID
```

Other examples are available from the AstroGrid web pages and from [19].

The Simple Image Access Protocol (SIAP) defines a prototype standard for retrieving image data from a variety of astronomical image repositories through a uniform interface. The IPHAS images are available through a SIAP [40] interface. The result of a query is, given a box centred at coordinates RA, Dec and a box size in degrees, a table of image CCDs which overlap the defined box. The table contains links to the processed image file itself. Alternatively the same list of images can be obtained with an appropriate SQL query using the *Query Builder*. At the moment only full image retrieval is supported. Although the graphical interface does not allow for a list of positions this is possible to accomplish from the command line interface using Python.

8.1.2 The Hipparcos Catalogue

A new reduction of the astrometric data as produced by the Hipparcos mission has been published [23], giving accuracies for nearly all stars brighter than magnitude Hp=8 to be better by up to a factor 4 than in the original catalogue [16]. This new astrometric catalogue and several supplementary catalogues are accessible through the Virtual Observatory.

The catalogue has been published in the VO in a similar way as described in the IPHAS catalogue above, i.e. ingesting the relevant tables into a relational database system and using DSA to publish it in a VO compliant way. Figure 6 displays the distribution of observed sources.

As an example the following ADQL query returns the Hipparcos number and parallax columns from the main catalogue (which has been aliased as m) for objects with a parallax greater than 200 milliarcseconds.

```
SELECT m.HIP,m.Plx from maincat as m where m.Plx > 200.0
```

The Hipparcos catalogue, as the IPHAS catalogue, can be queried both using a cone search or using ADQL.

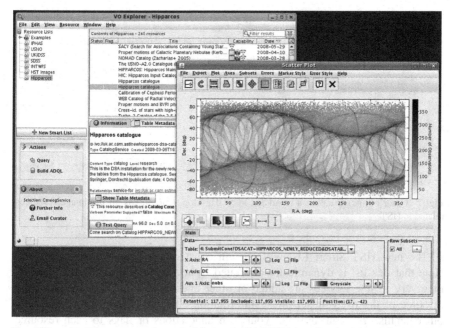

Fig. 6 Here the Hipparcos catalogue is "discovered in VOExplorer", selected and the result of a query against it shown displayed in TopCat. The distribution of sources in the Hipparcos catalogue colour coded against the number of observations per detection are plotted, which effectively demonstrates the Hipparcos scanning law

8.2 Science Use

Here we describe in detail some examples of utilising AstroGrid as an astronomer. In addition we note that a range of publications are now appearing which have made significant use of the AstroGrid VO system. For instance, mining of the SDSS and UKIDSS[38] catalogues in the search for Quasars [1], measuring the rate of solar magnetic flux emergence [11] and determining the properties of High-z radio starbursts [32]. These give an indication of the range of astronomy topics for which the VO can be used.

8.2.1 Stellar Jets

In this case we are going to search for HST images of stellar jets. First we are going to search the registry for resources matching HST which provide images. In order to do that first we need to create a list of resources that we want to query. In VOExplorer click on the New Smart List. Name your search "HST Images" and add the conditions as in the Fig. 7 (hint: clicking on the "+" icon adds another condition). When this is done click on Create.

---

[38] http://www.ukidss.org

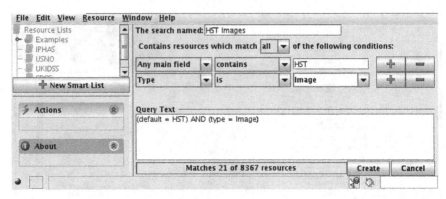

Fig. 7 (Color online) Creating a smart list in VODesktop which holds all the image resources related to HST

The new HST Smart List will contain a list of resources similar to Fig. 8. Note that this list is dynamically generated from a registry query so its contents may vary if new services are added or other ones are deleted. If you select one resource of the list you get in the Information tab a description of it and some other details. You can also add your own annotations or flags to each resource and even assign tags for easier search later. Each resource has a Curator or person in charge of it which can be emailed for clarifications or if the service is not working or not returning the expected results. The next step is to query all these resources for images. With the HST Smart List selected click on the menu Edit → Select All to select all the resources and then under Actions click on Query.

A new window will appear which is named VOScope. This is a generic interface which allows us to query the selected resources for images, catalogues and spectra. In our case we are only querying for images, so type HH30 in the Position or Object Name box (the name will be automatically resolved into coordinates) and 0.01 in the Search Radius (meaning 0.01 deg). Then click on the Search button. The selected HST resources are now being queried for images containing the HH30 object. A graphical display will appear similar to the one on Fig. 9

Double click on the CADC/HST Image Search and under actions Send Tables to Aladin. If you do not have that option available then it is a good time to start the Aladin browser now. The table we are sending contains the references to the actual images not the images themselves so Aladin displays a list of image files grouped by filter name. Just select two as click Submit (Fig. 10).

The procedure outlined above is fine if one wants to look at a particular object or two, but if we want to do this for several objects and also save the results then it is better to write a script.

The selected HST archive (defined in line 26) will now be searched for images for each object. A VOTable with the list of images for each object will be saved to the "hst" directory and also sent to TopCat.

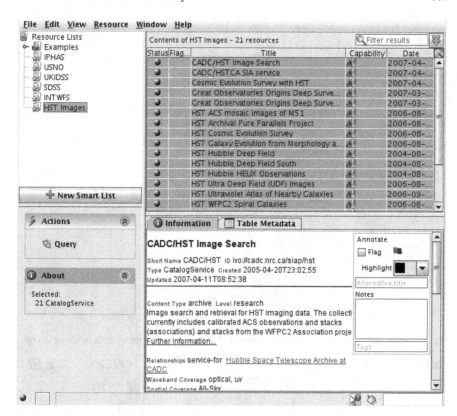

Fig. 8 (Color online) Newly created smart list with HST image resources

```python
#!/usr/bin/python

import os, time
from optparse import OptionParser
from astrogrid import acr, sesame
from astrogrid import SiapSearch

# Read command line arguments
parser = OptionParser()
parser.add_option("-b", "--broadcast", action='store_true',
                  default=False)
(options, args) = parser.parse_args()

# If we are going to send the tables to TopCat
# then start the plastic hub
if options.broadcast:
        acr.startplastic()
        time.sleep(2)

# Define list of objects
objects=['HH111', 'HH30', 'HH211', 'HH47', 'M16', 'Trifid',
         'HH524', 'Sigma Orionis', 'DG Tau', 'HL Tau',
         'M16', 'IRAS 04302+2247']
```

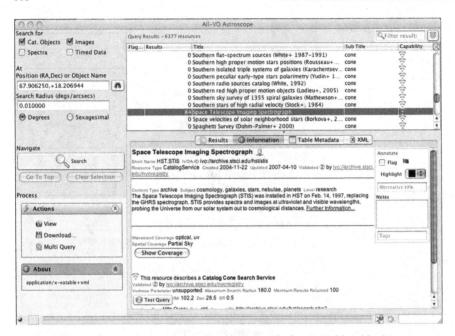

Fig. 9 (Color online) AstroScope window showing the results from a multi-archive image query

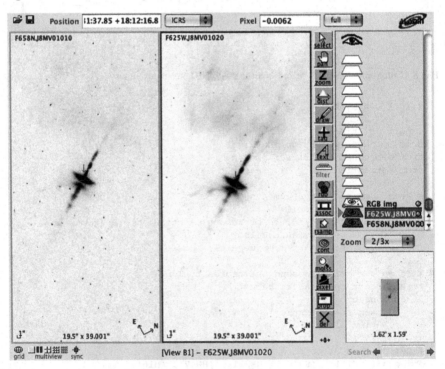

Fig. 10 (Color online) Images of jet HH30 displayed in Aladin

```
# Define the name resolver service
s = sesame()


# Define the endpoint of the search service
siap = SiapSearch('ivo://cadc.nrc.ca/siap/hst')

# Create output directory if it does not exist
if not os.path.isdir('hst'):
        os.mkdir('hst')

# Loop for each object
for obj in objects:
        coords, ra, dec = s.resolve(obj)
        ofile = 'hst/%s.vot' % obj.replace(' ','_').replace('+','p')
        if not os.access(ofile, os.F_OK):
                res = siap.execute(ra, dec, 30.0/3600.0, 0.01)
                open(ofile,'w').write(res)
        if options.broadcast:
                acr.plastic.broadcast(ofile, 'TOPCAT')
```

Once all image lists are loaded into TopCat it is possible to visualize the images in Aladin.

8.2.2 SDSS Field of Streams

A simple colour cut in the SDSS catalogue ($g - r < 0.4$) reveals the tidal stream of the Sagittarius dwarf spheroidal galaxy, as well as a number of other stellar structures in the field [8].

We can create a colour figure showing this results using AstroGrid, in particular, using the SDSS catalogue access available from the Query Builder. Figure 11 shows an example query which selects all those point-like objects in the SDSS catalogue with the above limit in the $g - r$ colour. Instead of sending a whole catalogue query to the archive we have selected to perform the task in 2 degree strips thus sending multiple queries and then joining back the result.

In order to do the full range from 0 deg to 60 deg we need to send then 30 similar queries; the only difference between them being the starting and ending declination values. This is a typical example of a case that makes scripting useful. It is relatively straightforward to create a sript which sends these 30 queries sequentially to the database and returns the resultant tables – an example of this is given in the AstroGrid web pages. After this is done all the tables are merged together using STILTS and loaded into TopCat for display. Using a colour code as in [8], i.e. blue for objects with $20.0 < r \leq 20.66$, green for objects with $20.66 < r \leq 21.33$ and red for objects with $21.33 < r \leq 22.0$ (corresponding roughly to a distance scale), we produce the plot in Fig. 12 comparable to Fig. 1 in [8]:

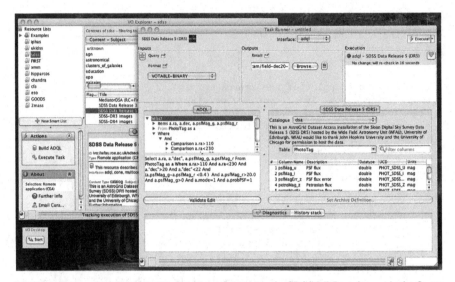

Fig. 11 (Color online) Example showing the query sent to the SDSS DR5 catalogue via the Query Runner tool interface – reached through VOExplorer

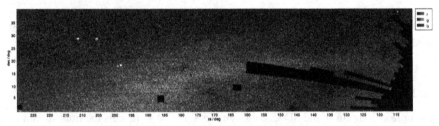

Fig. 12 (Color online) Spatial density of SDSS stars with colour $g - r < 0.4$. Effectively the colour represents distance, note clear evidence for structure and large scale streams

9 AstroGrid for the Astronomer

The AstroGrid system is available for download[39] from the main web-page at http://www.astrogrid.org. The only requirement is to have installed Java from Sun (version 1.5 or greater) which is available for a large variety of platforms including Microsoft Windows and Vista,[40] Linux[41] and MacOS.[42] If one wants to carry out some scripting then Python should also be installed which again is available for many platforms.

The AstroGrid website at http://www.astrogrid.org has a complete set of user documentation.

---

[39] http://www.astrogrid.org/wiki/Install/Downloads

[40] http://www.microsoft.com

[41] e.g. Ubuntu - see http://www.ubuntu.com)

[42] http://www.apple.com/macosx/

9.1 AstroGrid Website User Manual and Usage Examples

The AstroGrid user manual section describes in detail the VODesktop user interface and the complete range of functionality available.[43] The reference help described above is completed with examples of usage of VODesktop to do science. These include simple data access to do cone searches to catalogues, image query and retrieval, remote access to databases using ADQL (including, e.g., access to UKIDSS [13]), examples of catalogue cross matches, etc.

9.2 AstroGrid for the Data Publisher

Data and resource providers derive a number of significant benefits from their utilisation of the AstroGrid publishing infrastructure. First, there are significant economies in data distribution: reduced overheads in terms of user interfaces, access control, authorisation and interface protocols. Further, there is a single point of contact for information on Application Program Interfaces (APIs). Astro-Grid also offers flexible VO components, which enable the deployment of science data-specific algorithms, work-flows and processes and give the ability to rapidly create value-added data collections (i.e. science data plus associated data sets). Thus, AstroGrid offers interfaces to the VO for new missions that are developing their data-flow systems, which enables publishing their data products to the VO.

Those wishing to publish data to the VO making use of the AstroGrid, and thus Euro-VO, software system can find full details at http://deployer.astrogrid.org. This gives relevant links to related Euro-VO partner software such as the ESA DAL-toolkit [7] which can be used to publish image and spectra and is a natural complement to the AstroGrid DSA component used for publishing tabular and database held data.

10 Conclusions and Outlook

This chapter has given a brief overview of the Virtual Observatory. It has described in some detail the UK AstroGrid Virtual Observatory implementation, showing, though the use of a range of science use cases, how the AstroGrid software can be used by astronomers now. New users in the UK are welcome to find out more by visiting the AstroGrid web site.[44] From 2009, AstroGrid will be changing the focus of its activities to support more fully the use and deployment of its VO infrastructure by UK and Eruopean missions and facilities. This will be carried out under the

---

[43] http://www.astrogrid.org/wiki/Help/ReferenceHelp

[44] http://www.astrogrid.org

auspices of the new VOTC:UK project, thus also look there (http://www.votcuk.org for details.

For the wider European community, the AstroGrid software is being deployed and utilised by the Euro-VO. For full details the reader is encouraged to visit http://www.euro-vo.org.

Acknowledgements The authors would like to acknowledge the entire development and science team of the AstroGrid project who have been instrumental in creating the AstroGrid VO system.

The AstroGrid project is funded in the UK by the Science and Technology Facilities Council (STFC) with additional funding through its participation in the EU VOTECH, Euro-VO DCA (project RI031675) and AIDA (project RI2121104) activities.

The authors thank the organisers of the JETSET 5th School on High Performance Computing in Astrophysics, held at the Centre for Astronomy, National University of Ireland Galway, January 8–13 2008. This chapter amplifies the lecture and demonstration of the VO given at the School.

References

1. D'Abrusco, R., Longo, G., Walton, N.A., 2008, MNRAS in press (a-ph/0805.0156)
2. Adelman-McCarthy Adelman-McCarthy, J., Agueros, M.A., Allam, S.S., et al., 2007, ApJS 172, 634
3. Anderson, M. C., Jones, T. W., Rudnick, L., et al., 1994, ApJ 421, L31
4. *AstroGrid Phase A Report*,The AstroGrid Project, Oct 2003, available from: http://wiki.astrogrid.org/pub/Astrogrid/PhaseBDocs/redbook.pdf
5. Astronet *Existing national andregional strategic astronomy science plans*, http://astronet.sc.eso.org/web/ Cited 25 Jul 2008
6. *A Science Vision forEuropean Astronomy*, Astronet, Eds.P.T. deZeeuw, F.J. Molster, ISBN 978-3-923524-62-4http://www.eso.org/public/outreach/press-rel/pr-2007/Astronet_ScienceVision.pdf Cited 25 Jul 2008
7. Barbarisi, I., Salgado, J., Ortiz, I, 2007, Proc ADASS XVI, ASP Conf Ser 376, 611
8. Belokurov, V., Zucker, D.B., Evans, N.W., et al., 2006, MNRAS 642, 137
9. Bertin, E., Arnouts, S., 1996, A&AS 117,393
10. Bonnarel, F., Fernique, P., Bienayme, O., Egret, D., et al., 2000, A&AS 143, 33
11. Dalla, S., Fletcher, L., Walton, N.A., 2008, A&A, 479, 1
12. DeLaney, T., Rudnick, L., 2003, ApJ 589, 818
13. Dye S., Warren, S.j., Hambly, N.C., et al., 2006, MNRAS 372, 1227
14. Drew J.E., Greimel, R., Irwin, M.J., et al., 2005, MNRAS 362, 753
15. Ennis, J.A., Rudnick, L., Reach, W.T., et al. 2006, ApJ 652, 376
16. ESA, 1997, *The Hipparcos and Tycho Catalogues*, ESA SP-1200
17. Evans D.W., Irwin M.J., Helmer L., 2002, A&A 395, 347
18. Fesen, R.A., Morse, J.A., Chevalier, R.A., et al., 2001, AJ 122, 2644
19. González-Solares E. A., Walton, N.A., Greimel, R., et al., 2008, MNRAS 388, 89
20. Harrison, P., Winstanley, N., Taylor, J.D., Proc ADASS XVI, Proc Ser ASP 347, 291
21. *The International Virtual Observatory Alliance* http://www.ivoa.net Cited 25 Jul 2008
22. *IVOA Document Standards*, v1. 10,Working Draft 6 Feb 2008, http://www.ivoa.net/Documents/latest/DocStd.html Cited 24 Jul 2008
23. van Leeuwen F., 2007, A&A 474, 653
24. *Microsoft SQL Server*, http://www.microsoft.com/SQL/default.mspx
25. *MySQL*, http://www.mysql.com/
26. *The National Virtual Observatory*, main website, http://www.us-vo.org Cited 25 Jul 2008
27. Ortiz I., et al., 2008, http://www.ivoa.net/Documents/latest/ADQL.html

28. Perryman, M.A.C., Lindegren, L, Kovalevsky, J., et al., 1997, A&A 323, L49
29. *PostreSQL* www.postgresql.org
30. Quinn, P.J., Gorski, K.M., *Toward an International Virtual Observatory* ESO ASTRO-PHYSICS SYMP. ISBN 3-540-21001-6. Springer-Verlag Berlin/Heidelberg, 2004
31. Rajasekar, A., Wan, M., Moore, R., Schroeder, W., 2006, *A Prototype Rule-based Distributed Data Management System Rajasekar*, Proc HPDC workshop on "Next Generation Distributed Data Management", Paris, France.
32. Richards, A.M.S., Muxlow, T.W.B., Beswick, R., et al., 2007, A&A 472, 805
33. Rixon, G.T., Benson, K., Morris, D., Proc ADASS XVI, ASP Conf Ser 347, 316
34. Schlegel D.J., Finkbeiner D.P., Davis M., 1998, ApJ 500, 525
35. Skrutskie M.F., Cutri, R.M., Stiening, R., et al., 2006, AJ 131, 1163
36. *Sybase* http://www.sybase.com/
37. *Taverna Website* http://taverna.sourceforge.net/ Cited 25 Jul 2008
38. Taylor, J.D, Boch, T., Comparato, M., et al., 2007, Proc ADASS XVi, ASP Conf Ser 376, 511
39. Taylor, M.B., 2005, Proc ADASS XIV, ASP Conf Ser 347, 29
40. Tody, D., Plante, R., 2004, http://www.ivoa.net/Documents/latest/SIA.html
41. Walton, N.A., 2005, A&G 46c, 23
42. Walton, N.A., *VO-Tech Science Framework Document*, VOTECH Science Team, Oct 2008, available from: http://wiki.eurovotech.org/twiki/bin/viewfile/VOTech/VotcSFD?rev=1; filename=sfd-v1.0.pdf
43. Walton, N.A., Gonzalez-Solares, E., Allen, M.G., et al, 2006, Proc ADASS XV, ASP Conf Ser 351, 775
44. Walton, N.A., Witherick, D.K., Oinn, T., Benson, K.M., 2008, Proc ADASS XVII, ASP Conf Ser in press
45. Wikipedia, The Free Encyclopedia, *List of astronomical observatories*, http://en.wikipedia.org/wiki/List_of_observatories. Cited 23 Jul 2008
46. *Word Wide Web Consortium Process Document* http://www.w3.org/2005/10/Process-20051014/
47. York, D.G., Adelman, J., Anderson, J.E., et al. 2000, AJ 120, 1579

Part II
Applications in Astrophysics

Three Dimensional Continuum Radiative Transfer

Jürgen Steinacker

Abstract Radiative transfer is introduced as one of the grand challenge problems in astrophysics due to its key role of radiation as the main carrier of information and the high dimensionality of the problem. The relation to line RT is outlined and general solution methods are reviewed. We give the equation system for the stationary case of continuum radiation and discuss the different equation parts. As example for solution strategies, we discuss adaptively defined grids and ray-tracing algorithms dealing with high optical depth. The results of a two dimensional continuum RT benchmark are shown. Typical applications for continuum radiative transfer in star formation are presented in models for: (i) an evolving molecular cloud core as seen in SPH simulations and analyzed by forward RT calculations, (ii) SO-1 as largest circumstellar disk known so far, (iii) UC-1 as the first hypercompact HII region with a circumstellar disk candidate, (iv) IRS 15 as the first candidate for a remnant disk around a massive star, and (v) Rho Oph D where the inverse transfer modeling has led to the three dimensional density and dust temperature structure of the molecular cloud core.

1 Introduction

A high percentage of the information we are collecting from astrophysical sources is obtained by analyzing the radiation we receive from the objects. Therefore, radiative transfer (hereafter RT) is one of the most fundamental processes in astrophysics, and crucial for our interpretation whenever radiation is altered on its way from the source to the telescope. The alteration can range from small perturbations like the reddening of stellar light due to interstellar extinction or the change of polarization due to the Faraday effect, to almost complete shielding at short wavelengths and thermal re-emission at infrared and sub-millimeter wavelengths in the case of deeply embedded star-forming regions, envelopes, evolved stars, or tori around active galactic nuclei.

J. Steinacker (✉)
Max-Planck-Institut für Astronomie, Königstuhl 17, D-69117 Heidelberg, Germany, and
Astronomisches Recheninstitut am Zentrum für Astronomie Heidelberg, Mönchhofstr. 12-14,
D-69120 Heidelberg, Germany, stein@mpia.de

Steinacker, J.: *Three Dimensional Continuum Radiative Transfer*. Lect. Notes Phys. **791**, 117–135
(2009)
DOI 10.1007/978-3-642-03370-4_4   © Springer-Verlag Berlin Heidelberg 2009

1.1 Three Dimensional RT as a High-Dimensional Problem

By taking a closer look at the physics of RT, one can understand the relatively low number of existing three dimensional RT codes treating this challenge problem of computational astrophysics. Three-dimensional (3D) RT is a problem incorporating variables for wavelength, three spatial, two directional coordinates, and time. With a decent resolution of 100 points in each variable, RT calculations lead to solution vectors with 10^{14} entries. Beside this enormous requirement for the internal memory size of the used computer, the RT equation is an integro-differential equation including a scattering integral, making it difficult to apply common solvers. Hence, both the memory requirement and the complicated transfer solver make three dimensional RT a problem much more time-consuming than, e.g., hydrodynamic calculations including time and three spatial variables.

It may be noted that 100 grid points are often not sufficient to resolve all spatial structures, e.g., in proto-planetary accretion disks. They typically extend radially from hundreds of AU down to a few stellar radii, covering six orders of magnitude in density.

Also for the discretization of the direction, 100 grid points may not be enough to resolve the strongly peaked UV radiation of the stars. Steinacker et al. [23] calculated optimized equally spaced direction grid points on the unit sphere, but still 100×100 directional grid points correspond to a mean resolution of about $2.7°$ only.

Given the challenging character of the three dimensional problem, approximations concerning the spatial dimensions, the degree of complexity in the radiation field, or the physical properties of the medium causing the alteration of the radiation are commonly made.

Especially the structure approximations are a severe source of physical misinterpretation. They might be valid for shell structures or smooth disks, but most objects, e.g., in star formation regions or nearby galaxies show a substantial deviation from any symmetry once they are observed with sufficient resolution. This trend will continue with the oncoming updated or new interferometers such as PdB, VLTI, and ALMA reaching resolutions revealing details at milli-arcsecond level. Therefore, there is ample need to have three dimensional RT codes available and tested from the observational side.

From the simulation view point, three dimensional Smooth-Particle Hydrodynamical (SPH) and Magneto-Hydrodynamical (MHD) simulations are now common and produce time-dependent density and temperature distributions of dust and gas for many galactic or extragalactic applications. Without a three dimensional RT code, it is impossible to predict at which wavelengths or by which telescope these three dimensional features can be detected. Moreover, due to the higher dimensionality, in three dimensional MHD simulations RT is treated in crude approximations ranging from ignoring energy transport by radiation, flux-limited diffusion, to ray-tracing the radiation from the source to the cells. A verification of the validity of the approximation is possible only by using a full three dimensional RT code.

1.2 Comparing Continuum and Line RT

For the physical interpretation of observed images, often both sets of line and continuum data have to be considered simultaneously.

Line RT generally has to be performed for atomic or molecular effects of photon absorption, scattering, or emission. As the atoms and molecules are moving and the lines appear in a narrow wavelengths range, the emission is strongly influenced by the velocity field within the astrophysical structure to investigate. Furthermore, the abundance of the atom or molecule within the gas has to be known often depending on a complex network of chemical reactions. This has several effects on calculating and modeling with line RT. First, parameter numbers of the order of several hundreds are needed to describe the model of the gas in its morphology, chemical composition, and kinematics. As the images and SEDs contain line-of-sight-integrated information only, modeling line data means to deal with possible model ambiguities. The parameters are also strongly coupled so that it is hard to disentangle them just from the line data. On the other side, there is hope the richness of the data will enable us to improve our understanding of the astrophysical processes substantially once three dimensional line transfer codes are able to determine all parameters. Second, to know the emission of a cell of gas, the level population of the molecule needs to be determined. Given the often complex level structure of the important molecules, this will slow down the line RT calculation (e.g., compared to continuum codes by a factor 10–100). Third, a derivative term with respect to wavelengths appears in the equation system that is not present in the continuum case.

Continuum RT is considered, e.g., for dust particles as small solid bodies which absorb, scatter, and re-emit radiation. The micron-sized particles that commonly dominate the RT are well-mixed in the gas, but they can follow a size and shape distribution [17] as well as a distribution of chemical compositions. Smaller particles down to a size of a few atomic layers do not emit as black bodies, and their time-dependent re-emission of photons again requires more computational effort to calculate the local mean dust temperature [13, 18].

Here, we will concentrate on the simple case of continuum RT of dust particles which have the size of typical interstellar dust particles, so that they can be described as black bodies.

1.3 Solution Methods

Each of the solution algorithms used so far has its advantages and drawbacks.

In *Monte Carlo methods*, a photon is propagated through the calculation domain and its scattering, absorption, and re-emission is tracked in detail [33, 35, 12, 15, 21]. This allows to treat very complicated spatial distributions [22], arbitrary scattering functions [5], and polarization [32, 3]. Monte Carlo methods encountered difficulties when covering re-emission in all directions over many events, and for very small or very high optical depths [16]. Meanwhile, several algorithms have been proposed to deal with high optical depths. Adaptive ray-tracing can be applied to the direction space [1]. The major drawback, however, is that there is no global error control when using Monte Carlo schemes.

Ray-tracing solvers can treat arbitrary density configuration with full error control. General purpose solvers for ordinary differential equations can be used to overcome the problem of strongly varying optical depth when using the ray-tracing solution method. These schemes are available in high-order accuracy and with adaptive stepsize control (e.g., advanced fifth-order Runge Kutta solvers). But they require the implicit re-calculation of the stepsize and become very time-consuming when the optical depth is high, so that more sophisticated solvers are required.

Grid-based solvers, in combination either with finite differencing or short characteristics, have the advantage of error control on the grid and a drawback is the stiff grid so that resolution of complex three dimensional structures needs an appropriate grid generation algorithm [27]. Adaptive grids became standard in lower-dimensional problems like hydrodynamical calculations – for RT, the refinement criterion is less clear. An interpolation between the grids for the temperature iteration is numerically costly though and gives rise to interpolation errors in the obtained solution. In most papers presenting grid-based three dimensional RT codes, numerical diffusion has not been considered and taken into account [24]. The effect is well-known in the course of discretizing hyperbolic equations.

Moment methods are well-posed to treat the optical thick regime and are usually applied to radiation fields with a moderately varying direction dependency. They encounter problems when describing a strongly peaked radiation field arising in the parts of the computational domain where the optical depth is at the order of unity.

An overview on existing continuum RT codes across different astrophysical fields can be obtained from the webpage of a recent radiative transfer workshop www.mpia.de/RT08.

2 Stationary Three Dimensional Continuum RT: Equation and Solution Strategy Examples

2.1 The Stationary Three Dimensional Continuum RT Equation

We describe the stationary radiation field by the total specific intensity $\mathcal{I}_\lambda(\lambda, \mathbf{x}, \mathbf{n})$, where $\mathbf{x}$ gives the location in space, $\mathbf{n}$ is the direction of the radiation, and λ its wavelength. Starting with the boundary values, we can calculate the transport of the radiation through the considered domain by solving the stationary three dimensional continuum RT equation

$$
\mathbf{n}\nabla_{\mathbf{x}}\mathcal{I}(\lambda, \mathbf{x}, \mathbf{n}) = -\left[\kappa^{abs}(\lambda, \mathbf{x}) + \kappa^{sca}(\lambda, \mathbf{x})\right]\mathcal{I}(\lambda, \mathbf{x}, \mathbf{n}) + \kappa^{abs}(\lambda, \mathbf{x})\, B[\lambda, T(\mathbf{x})]
$$

$$
+ \frac{\kappa^{sca}(\lambda, \mathbf{x})}{4\pi}\int_\Omega d\Omega'\, p(\lambda, \mathbf{n}, \mathbf{n}')\,\mathcal{I}(\lambda, \mathbf{x}, \mathbf{n}') \tag{1}
$$

with the Planck function B, the dust temperature T, and the phase function $p(\lambda, \mathbf{n}, \mathbf{n}')$ giving the probability that radiation is scattered from the direction $\mathbf{n}'$ into $\mathbf{n}$, with the solid angle Ω'. For simplicity, we have skip the λ index of the intensity which indicates that it is defined per wavelength interval. We will restrict our consideration to the use of dust particles with one size and a specific chemical composition. Equation (1) can also be used for a size or composition distribution of dust grains, but then each of the different dust species will have its own temperature.

The physical quantities describing the efficiency to absorb and scatter the incident radiation by an ensemble of dust grains can be written as

$$\kappa^{abs,sca}(\lambda, \mathbf{x}) = k^{abs,sca}(\lambda, \mathbf{x})\,\rho(\mathbf{x}), \qquad (2)$$

where $\rho(\mathbf{x})$ is the dust density and $k^{abs,sca}(\lambda, \mathbf{x})$ are the mass absorption and scattering coefficients of the dust ensemble, respectively.

The dust particle cross sections for absorption and scattering are shown in Fig. 1 for a 0.12 micron-sized dust particle [9]. A variety of optical properties of dust particles can be found at the Jena-Petersburg database www.astro.uni-jena.de/Laboratory/Database/databases.html.

Two source terms of radiation are explicitly given in Equation (1): scattering into the line of sight and re-radiation by the dust particles.

Intensity and dust temperature are not independent. The radiation field determines the temperature and in turn the dust re-emission contributes to the radiation field. This couples the partial integro-differential RT equation to the local energy balance equation describing how a dust particle is heated by the source radiation and the radiation of all other particles. The balance equation for the energy density in local thermal equilibrium at point $\mathbf{x}$ is

$$\int_0^\infty d\lambda\; Q^{abs}(\lambda)\, B[\lambda, T_{rad}(\mathbf{x})] = \int_0^\infty d\lambda\; Q^{abs}(\lambda)\, \frac{1}{4\pi} \int_\Omega d\Omega'\, \mathcal{I}(\lambda, \mathbf{x}, \mathbf{n}'). \qquad (3)$$

The temperature is denoted by T_{rad} to distinguish it from temperatures arising from other possible heating sources like viscous heating, cosmic rays, or gas–grain collisions.

A simultaneous treatment of the three dimensional RT and MHD equations in one code is currently beyond the capabilities of nowadays computers. MHD simulations commonly use an approximate RT to calculate radiative heating, while in turn RT codes can use the derived densities and the heating sources to calculate the radiation field at a given time or in a stationary picture.

In view of the substantial computational effort to solve the three dimensional transport equation, it is mandatory to use any approximation that is allowed by the physical conditions. With vanishing scattering integral, (1) becomes a first-order differential equation which can be solved without problems using ray-tracing. To make use of this approximation, and since the operators in the integro-differential equation (1) are linear in $\mathcal{I}$, we can (following, e.g., [11]) split the total specific intensity into an unprocessed passing source component I^* that includes also the thermal contribution from the dust, and a processed component I of radiation that has encountered scattering

$$\mathcal{I} = I^* + I. \tag{4}$$

Substituting the source term to be

$$S(\lambda, \mathbf{x}, \mathbf{n}) \equiv \frac{1}{4\pi} \kappa^{sca}(\lambda, \mathbf{x}) \int_\Omega d\Omega'\, p(\lambda, \mathbf{n}, \mathbf{n}')\, I(\lambda, \mathbf{x}, \mathbf{n}') + C^*(\lambda, \mathbf{x}, \mathbf{n}), \tag{5}$$

this leads to three equations

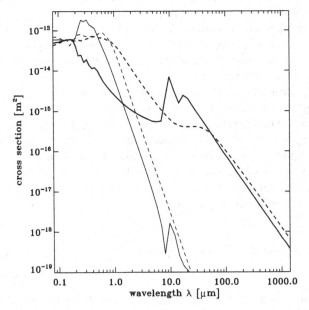

Fig. 1 Cross sections of a 0.12 micron-sized spherical homogeneous dust particle for absorption (*thick line*) and scattering (*thin line*) consisting of silicate (*solid*) or carbon (*dashed*)

$$\mathbf{n}\nabla_{\mathbf{x}} I^*(\lambda, \mathbf{x}, \mathbf{n}) = -\kappa^{ext}(\lambda, \mathbf{x})\, I^*(\lambda, \mathbf{x}, \mathbf{n}) + \kappa^{abs}(\lambda, \mathbf{x})\, B[\lambda, T(\mathbf{x})] \tag{6}$$

$$\mathbf{n}\nabla_{\mathbf{x}} I(\lambda, \mathbf{x}, \mathbf{n}) = -\kappa^{ext}(\lambda, \mathbf{x})\, I(\lambda, \mathbf{x}, \mathbf{n}) + S(\lambda, \mathbf{x}, \mathbf{n}). \tag{7}$$

$$\int_0^\infty d\lambda\, Q^{abs}(\lambda)\, B[\lambda, T(\mathbf{x})] = \int_0^\infty d\lambda\, Q^{abs}(\lambda) \frac{1}{4\pi} \int_\Omega d\Omega'\, I(\lambda, \mathbf{x}, \mathbf{n}') + D^*(\mathbf{x}) \tag{8}$$

with the known contributions

$$C^*(\lambda, \mathbf{x}, \mathbf{n}) = \frac{\kappa^{sca}(\lambda, \mathbf{x})}{4\pi} \int_\Omega d\Omega' \, p(\lambda, \mathbf{n}, \mathbf{n}') \, I^*(\lambda, \mathbf{x}, \mathbf{n}') \qquad (9)$$

$$D^*(\mathbf{x}) = \int_0^\infty d\lambda \, Q^{abs}(\lambda) \frac{1}{4\pi} \int_\Omega d\Omega' \, I^*(\lambda, \mathbf{x}, \mathbf{n}') \qquad (10)$$

and the abbreviation $\kappa^{ext} = \kappa^{abs} + \kappa^{sca}$. $C^*(\lambda, \mathbf{x}, \mathbf{n})$ represents the source and thermal radiation that is scattered at the point $\mathbf{x}$ into the direction $\mathbf{n}$. The source and thermal contribution to the heating is described by $D^*(\mathbf{x})$.

As intended by the split (4), the first transfer Equation (6) can be transformed to a path integral and thus easily be calculated using the formal exponential solution. For an empirical optical data set, the numerical solution is conveniently derived for all wavelengths, e.g., using fifth-order Runge Kutta with adaptive stepsize control as ray-tracing routine. This provides error control for the solution, and can be used for all optical depths from thin to thick regions. Moreover, it allows to treat multiple external and internal sources of radiation (Fig. 2).

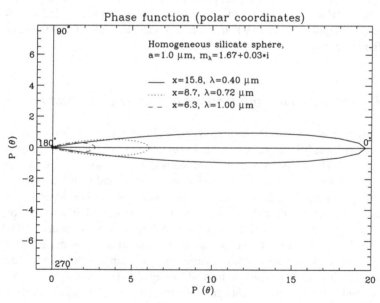

Fig. 2 (Color online) Scattering phase function in polar coordinates for homogeneous spherical silicate particle of size $a = 1\,\mu$m for different wavelengths. m_λ denotes the mean complex refractive index for all three plotted wavelengths and $x = 2\pi a/\lambda$ is the size parameter

The second Equation (7) still has an integro-differential form and requires a separate, more sophisticated treatment. The integral over the unit sphere can be done efficiently, e.g., by using direction grid nodes that are equally spaced over the unit sphere [23]. The third Equation (8) allows to update the temperature from an intensity that has been calculated assuming a fixed temperature using a standard accelerated Λ-iteration [10]. Alternatively, the idea of *immediate re-radiation* can be used to directly update the temperature information while solving the ray equation with a ray-tracer, following the Monte Carlo photons [2], or while performing finite-differencing calculation on a grid.

2.2 RT on Adaptively Refined Grids

With a finite-differencing scheme, the ray equation can be discretized spatially but the linear grids will not work to resolve the large density gradients. We have proposed to calculate an adaptively refined spatial grid for each wavelength separately as the optical depths strongly varies with wavelength [26]. Through minimizing the first-order discretization error in the scattered radiation intensity, we have provided global error control for solutions of RT problems on the grid. In order to reduce the grid point number in regions where the optical depth becomes large we have proposed the use of the concept of penetration depth. The proposed grid generation algorithm is easy to implement and allows pre-calculation of the grids and storage in integer arrays, making a fast solution of the three dimensional RT equation possible. The drawback of this method is the cost of interpolation calculations and the introduced interpolation errors when it comes to calculating the temperatures using the different grids.

2.3 Ray-Tracing Through Regions of Very High Optical Depth

Especially in massive star formation, the optical depth can vary by six orders of magnitude or more [4]. Commonly, most solvers, be they a Monte Carlo approach or ray-tracing, will be forced to perform small steps slowing down the code. A new ray-tracing technique has been proposed by [30] to use the optically thick approximation in order to speed up the calculations. Assuming radial power-law dependencies for the density and temperature distribution, we have calculated the absolute solution errors, the numerical effort, and the stepsize variation for a given accuracy. We have shown that advanced ordinary partial equation solvers like fifth-order Runge Kutta with adaptive stepsize control are too expensive to be applied to the inverse three dimensional RT modeling problem. Instead we suggested a second-order ray-tracing scheme controlling the relative change of the intensity and making use of the diffusion approximation in the regions of high optical depth. The method is designed to cross optically thick regions quickly, to resolve the important regions with an optical depth around unity, and to have a moderate computational expense mandatory for inverse three dimensional RT modeling. We apply the method to calculate a far-infrared image of a dense molecular cloud core (being the initial configuration of a star formation process) with speed-ups of the order of several hundreds. Figure 3 shows a physical setup that was chosen in order to include all possible cases where solvers encounter problems. The bottom right figure part depicts the error of the new solver (thin line) and the interpolation error (thick line). In the regions of high optical depth, relative errors around 0.1 per mil could be achieved.

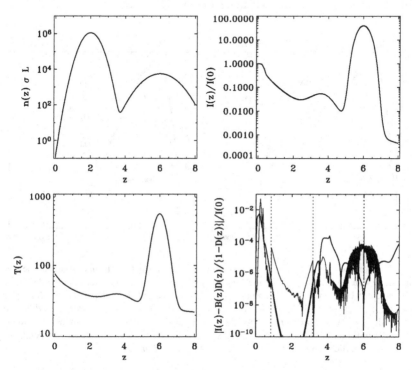

Fig. 3 Overview of the physical setup and solutions for the test-case ray-tracing calculation. *Top left*: Optical depth along the ray. *Bottom left*: Temperature along the ray. *Top right*: Intensity along the ray using a fifth-order Runge Kutta solver for the ray equation. *Bottom right*: Normalized difference between the solutions obtained from the accurate solver and the relative intensity ray tracer (*thin line*) and interpolation error (*thick line*). Regions where the solver uses the optically thick approximation are enclosed by vertical dotted lines ($z = 0.9$ to 3.2, and around $z = 6$)

2.4 RT Benchmark Results

As analytical solutions of RT problems are impossible except for a few cases treated in one dimensional [25], benchmark solutions are desired to compare the outcome of the different solvers for a well-tested problem. This has been done for the one dimensional case by [14]. For a two dimensional disk configuration, [20] have presented benchmark problems and solutions. In this paper, the reliability of three Monte Carlo and two grid-based codes was tested by comparing their results for a set of well-defined cases which differ for optical depth and viewing angle. The disk density was assumed to follow a simple parametrized form and was irradiated by a single central T Tauri star. For all the configurations, the overall shape of the resulting temperature and spectral energy distribution was well reproduced. Figure 4 shows the spectral energy distributions for varying inclinations and disk masses. The solutions can be used for the verification of other RT codes.

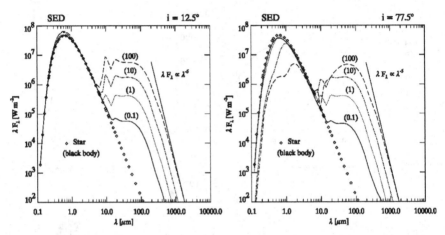

Fig. 4 SED for two disk inclinations i as given on top of each panel. Each *curve* provides the mean value of the five RT simulations for the four computed models with different optical depth. The midplane optical depth is given in parenthesis labeling each curve. In both panels *solid lines* show results for the most optically thin disk, *dotted lines* for a disk having $\tau_v = 1$, *dot-dashed lines* for a disk with $\tau_v = 10$ and *dashed lines* for the most optically thick model. Diamonds provide the black-body emission from the naked star. The slope of the SED at long wavelengths depends only on the dust properties and is plotted in each panel with a *solid line*

3 Applications

Given the wealth of application for RT in astrophysics, a complete overview over all aspects of RT calculations resulting from the different applications would be well beyond the scope of this contribution. Instead we illustrate the requirements and capabilities of RT using a few recent applications from the star formation research. Other important fields are, e.g., stellar atmospheres, late phases of stellar evolution, tori around active galactic nuclei, and quasars.

Star formation with the drastically varying optical depths, and the still unsolved question of what mechanism are controlling the onset and the first phases of the formation process is an exciting and rich area to apply RT codes.

3.1 Dust Absorption and Emission: An Evolving Molecular Cloud Core Seen in Space

To investigate the early phase of star formation, the results of three dimensional simulations have been investigated in [28]. In the simulations, it was assumed that initially low-mass condensations pass through a stage of turbulence-dominated condensation where they accumulate mass and merge together to form extended prestellar core-like objects. The typical density structures in the cores were non-spherical throughout their evolution. Using a three dimensional continuum RT program, we have generated images at 7, 15, 175 μm, and 1.3 mm for different evolutionary times and viewing angles. As an example, Fig. 5 shows images for an evolution time of 56,000 years after the onset of the gravitation at these wavelengths. We showed that projection effects can lead to a severe misinterpretation of images: A one dimensional analysis of the vicinity of the density maxima would suggest density profiles in agreement with one dimensional-core collapse models. The underlying density structure, however, is intrinsically three dimensional and deviates strongly from the obtained one dimensional model distribution.

3.2 Scattered and Absorbed Light from Massive Stars: Massive Disk Candidates as Indication for a Formation Via an Accretion Disk

While observations show that young low- and intermediate-mass stars are surrounded by a circumstellar disk which are often actively accreting matter from it, only a few candidates for such a circumstellar disk around a massive star are known (see [7] for a list of candidates). It was argued by [34] that the strong radiation pressure of the young massive star might hinder spherically symmetric accretion onto the star to reach high masses of several tens of solar masses. [36] have performed two dimensional hydrodynamical grid simulations of a collapsing cloud including non-gray RT with flux-limited diffusion. We found that for initial cloud masses of 120 $M_\odot$, a massive disk can form around the central star with final masses of the star reaching about 43 $M_\odot$. Alternatively, [6] proposed that merging of intermediate-mass stars may lead to the formation of massive stars. The detection of disks around young massive stars would provide a mean to investigate this question.

3.2.1 SO-1 – The Largest Circumstellar Disk Known So Far

[31] have analyzed the prominent silhouette structure in M 17 showing a symmetric large-scale pattern in absorption against a bright background, with a central emission region, an hourglass-shaped reflection nebula perpendicular to the extinction bar, a complex outflow over and below the dark extinction lane, and signatures for accretion of matter. Due to the large scale and the strong symmetry of the structure, as well as its presence within a massive star formation region, it attracts special

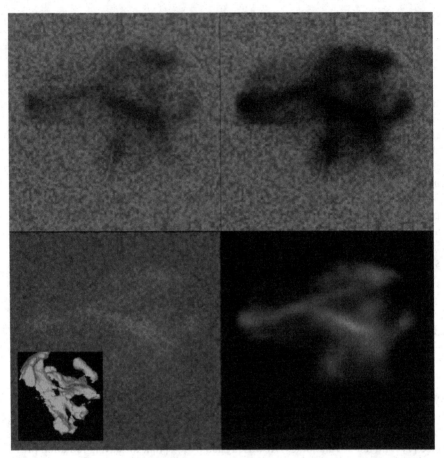

Fig. 5 (Color online) Images of the cloud core fragment at the wavelengths 7 (*top left*), 15, 175, and 1300 μm (*bottom right*), at a simulation time of 5.6×10^4 years after the start of the gravitational influence. The inset shows the iso-density surfaces of the used averaged SPH density distributions of the cloud core fragment corresponding to a density of 4×10^{-17} kg m^{-3}

attention as a candidate for a massive disk around a star that might be massive or has the potential to reach such a mass. While the estimate of the disk mass for most massive disk candidates comes from low resolution FIR/mm measurements, we used the advantage that due to the background illumination, the column density can be determined at $\lambda = 2.2$ μm from a high-resolution NAOS/CONICA image.

We investigated whether the observed extinction structure is consistent with a model of a circumstellar rotationally symmetric flattened density distribution. Applying a commonly used analytical disk model with a power law in radius and a vertical Gaussian distribution, we have fitted a seven parameter model to the few 8000 pixel of the image. The PSF of a point source was covered by about 3.3 pixels leading to about 2400 independent data points. We found that the derived optical depth is consistent with a rotationally symmetric distribution of gas and dust around

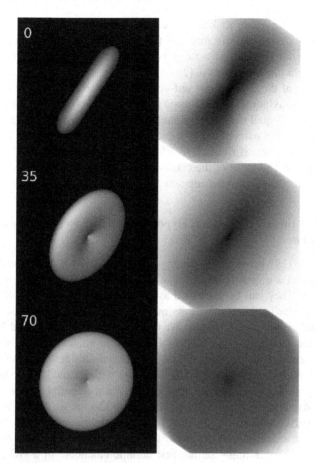

Fig. 6 Theoretical absorption image of the circumstellar disk using the optimized fit parameters and varying the inclination angle against the edge-on case. The *left panel* shows an iso-density surface of the disk distribution (the inclination angle against the edge-on case is given in degrees). The *right panel* shows the corresponding 2.2 μm absorption image

the central emission peak. The extent of the axisymmetric disk part is about 3000 AU, with a warped point-symmetrical extension beyond that radius, and therefore larger than any circumstellar disk detected so far. The resulting theoretical absorption image of the disk is shown in Fig. 6 for different inclination angles, along with the corresponding 2.2 μm-image.

Surprisingly, we find a flat radial density power law exponent of −1.1 indicating a density distribution with almost constant surface density. The large e-folding scale height ratio of about 0.5, however, questions the applicability of height-integrated quantities and is substantially larger than the e-folding scale heights derived for circumstellar disks around intermediate- and low-mass stars.

The mass of the entire disk estimated from the column density is discussed depending on the assumed distance and the dust model and ranges between 0.02 and 5 $M_\odot$. The derived disk mass range is smaller than the mass range of 110–330 $M_\odot$ derived from the 13CO data in [7].

The stability of the disk against self-gravitational forces is analyzed calculating the ratio of the gravitational acceleration by the central object and the disk for the interesting case of equal mass for star and disk. We find that the disk shows deviations only in the outer parts of the disk due to the large e-folding scale height factor of about 0.5. The disk distribution shows hints of point-symmetrical disturbances in the outer parts. The mass of the central object cannot be constrained rigorously from the existing data due to the edge-on configuration.

3.2.2 K-Band Images of the First Hypercompact HII Region with a Disk Candidate

Nielbock et al. [19] investigated the hypercompact (HC) H II region M17-UC1. HC H II regions are commonly associated with the earliest stages of high-mass star formation. They can be explained as a phase in early stellar evolution, when the high-mass star begins to ionize its molecular accretion flow. In such a stage, ionization and accretion can coexist. We have presented new NIR and MIR observations of such a hypercompact H II region. As a prominent feature, the Ks-band image shows a dark lane in scattered light producing a substantial silicate absorption feature observed in the MIR. Analyzing the image with RT models, we find the best agreement by assuming a disk-like structure around the central source instead of a foreground filament (the K-band image, the disk model image, as well as a filament model image are shown in Fig. 7). For this reason and because of the "high-mass" nature of the embedded star as determined by many independent measurements, we suggest that M17-UC1 might be the first definite candidate of a HC H II region where at least parts of a circumstellar disk are still present.

Fig. 7 Model fit of M17-UC1. *Left*: Original K-band image. *Middle*: Disk model inclined at $30^\circ irc$ as discussed in the text. *Right*: Scattered light image at 2.2 μm from a filament in front of a B0 star. The filament has a diameter of about 3000 AU and a distance of 5000 AU to the star; the dust properties are identical to those of the disk model

3.2.3 The First Candidate for a Remnant Disk Around a Massive Star

The latest stage of the formation of a massive star, i.e., a final massive star that has stopped accretion but that is still surrounded by circumstellar material left over from its birth, has escaped detection so far. If formed through accretion, a flat symmetric dusty disk should be visible before it becomes evaporated by the central hot star.

Chini et al. [8] argue to have found a good candidate for such an evolutionary stage, i.e., a massive star that has stopped accretion but that is still surrounded by an extended reservoir of remnant circumstellar dust. The new source IRS 15 is located in M17 at a distance of about 2.2 kpc and has been discovered as a visible star associated with an infrared excess. The multicolor imaging and spectroscopy study presents new physical properties of both the star and its circumstellar environment. The morphology of the dust distribution with a projected axis ratio of 1.32 at $10 \, \mu m$ is consistent with two extreme cases: an oblate envelope seen edge-on (with a minimal axis ratio of about 2 : 1) or a flat disk seen under an inclination angle of about 53° with respect to the line of sight. While a possible envelope of IRS 15 could be deformed on its northeastern side, facing the nearby OB star cluster, there is no obvious physical reason for its flatness on the other side. We therefore argue that the large-scale symmetry of the emission pattern around IRS 15 most likely arises from a circumstellar disk, although a flat envelope cannot be entirely excluded. The main axis of the ellipsoidal emission is perpendicular to the direction of the strong radiation field from the neighboring massive stars. It can be argued that their strong radiation field may have influenced the orientation of the disk, if not triggered the formation process of IRS 15. Moreover, slight distortions of the disk may arise

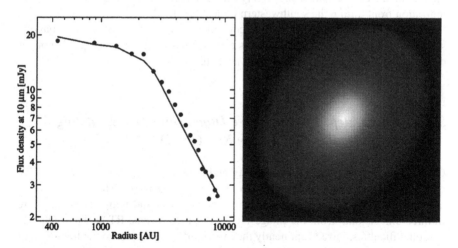

Fig. 8 Model of the infrared emission of IRS 15. *Left*: The observed radial $10 \, \mu m$ flux profile (as an example, the *filled circles* show a profile along the main axis) can be reproduced by a density distribution (*solid curve*), which displays a flattening in its inner part. *Right*: The resulting $10 \, \mu m$ model image; the field of view is $14'' \times 14''$

from an interaction of the radiation and stellar winds of the nearby O stars with the disk material.

To model the elliptical emission at 3.7 and 10 μm, we have used a thin disk with a scale height ratio of 0.1, an inner radius of 130 AU, an outer radius of 10,000 AU, and 0.1 μm-sized silicate dust particles. The best results of the model fits using optically thin ray-tracing through the disk can reproduce the observed radial flux profiles fairly well. We obtain a density distribution that displays a flattening in its inner part. Compared to standard accretion disk models, a substantial fraction of the inner disk must have been already eroded. This may indicate a similar mass removal process as observed for older disks around lower mass objects. Another argument in this direction comes from the K-band spectra where the CO band heads beyond 2.29 μm usually arising at the inner disk from hot and dense materialare missing. The model requires a total dust mass of $5 \times 10^{-4} M_\odot$. The measured flux density is compared to the model density in Fig. 8 (left) and the resulting model image is shown (right).

IRS 15 is very likely a star on the main sequence that has stopped accretion. This is corroborated by our spectra where typical accretion indicators like emission lines of Hα, the Ca II triplet λ8498, λ8542, and λ8662, and He I λ6678 are absent. On the other hand, IRS 15 must be extremely young because the time span during which a massive star can maintain a circumstellar disk that has not yet been completely destroyed by the strong stellar winds must be only a few mega-years. Since the remnant disk surrounds a massive, ionizing star and is also being exposed to external ionizing radiation, it should show evidence for ionized gas, in particular in the form of radio free-free emission. However, IRS 15 is located within the extended maximum of free-free emission in M17, which is interpreted as originating from the ionization front hitting the southwestern molecular cloud.

In summary, IRS 15 is a fortunate coincidence where the newly born massive star is already optically visible while its circumstellar disk (or envelope) is not yet dispersed and still glowing at infrared wavelengths.

3.3 The Next Step: Inverse Three Dimensional RT Modeling of the Molecular Cloud Core Rho Oph D

Calculating images from a given density and temperature distribution using a RT code can quickly reach the capabilities of current computers. The finally desired approach would be, however, to determine the density and temperature structure in three dimensional from the images. This so-called *inverse* RT problem features several difficulties, most prominently the loss of information due to the line-of-sight integration performed during the observations. It was shown in [29] that inverse RT transfer is possible nevertheless already for the analysis of molecular cloud cores. We proposed a new method to model the three dimensional dust density and temperature structure of such cores. The method is based on the fits of multiple continuum images in which the core is seen in absorption and emission and requires only a few computations with a three dimensional CRT code. The large parameter space of a three dimensional structure is covered using simulated annealing as optimization algorithm. One of the key points of the proposed method is the use of a $T_{\bar{\tau}}$-relation that has been derived from the three dimensional CRT runs. It links the mean optical depth at a given wavelength from outside to a spatial point within the core to the dust temperature at this point, making a fast evaluation of the mm emission along the line of sight possible.

We have applied the method to model the dense molecular cloud core ρ Oph D. In the MIR, the core is seen in absorption against a bright background from the photo-dissociation region of the nearby B2V star, illuminating the cloud from behind. Two ISOCAM images at 7 and 15 μm have been fitted simultaneously by representing the dust distribution in the core with a series of three dimensional Gaussian density profiles. The background emission behind the core was interpolated from nearby regions of low extinction. Using simulated annealing, we have obtained a two dimensional column density map of the core. The column density of the core has a complex elongated pattern with two peaks, with the southern peak being more compact. To retrieve the full three dimensional structural information, we have calculated the temperature structure of the core with a three dimensional CRT code, assuming several limiting cases of core extent along the line of sight. It was shown that, for a given position in the core, the relation between the dust temperature and the mean optical depth from outside to this position varies within less than 1 K, when changing the shape of the core along the line of sight. Therefore, the $T_{\bar{\tau}}$-relation was used to estimate quickly the temperature at a given point by a fast direction averaging over the optical depth at just one wavelength. Using this mean, we have varied extent and position of the Gaussian components of the density along the line of sight until a fit of both the MIR and a 1.3-mm map, obtained with the IRAM 30 m telescope, could be achieved. We have presented the three dimensional dust density and temperature structure, revealing a condensed southern part and a more extended and complex northern part. We have addressed several sources of errors, namely the background determination method, the assumed dust particle properties, errors in the maps, the ambiguity in the derived distribution, and the approximation error of the $T_{\bar{\tau}}$ method. As we did not reach a perfect fit of the mm image between the two maxima due to the low number of Gaussians, we expect that the detailed three dimensional structure in-between the density maxima might change when using a higher number of Gaussians. The general structure of a condensed southern maximum and a complex northern multi-clump region, however, is clearly resolved within the limits of the presented analysis. This structure is in general agreement with recent gravo-turbulent collapse calculations for molecular clouds. We speculate that the southern condensation belongs to the category of cloud clumps which are dominated by self-gravity. It may be collapsing to form a star in the near future. The northern clumps, however, may still have sufficiently large internal or kinetic energy to re-expand and merge into the ambient medium once the turbulent compression subsides. Figure 9 illustrates the resulting three dimensional density data cube by showing two iso-density layers (semi-transparent and solid).

Fig. 9 (Color online) Two layers of constant density (*solid and semi-transparent*) illustrating the full three dimensional configuration of the molecular cloud core ρ Oph D. The density was determined fitting a continuum map at 1.3 mm and two continuum maps at 7 and 15 μm

In this application, we determine about 210 free parameter to fit about 2000 pixels simultaneously, but hidden in the assumptions are many more free parameter like the dust properties or details of the illuminating background. It has to be kept in mind, though, that the hidden parameters are present in any other, more simple modeling as well, and often with more unrealistic approximations. The ρ Oph cloud is illuminated from behind in contradiction to a one dimensional boundary condition for the incoming radiation.

The ultimate goal of applying the method to well-observed cores, however, will be to address the key question of early star formation, namely if the considered cores have in-falling material. The current line observations provide the molecular line emission flux integrated over all moving gas cells along the line of sight. In the general case, gas motion and emissivity of the cells can not be disentangled, and the one dimensional approximation or shearing layers are assumed to unfold it. Without unfolding it, infall motion can be mixed up with rotational motion leaving it undecided if the core shows any sign for the onset of star formation. This is changed if the core has been investigated with the $T_{\bar{\tau}}$-method. Knowing the full three dimensional structure in dust density and temperature, the line of sight-integral can be inverted providing the complete kinematical information if the considered line is optically thin and a model for the depletion of the considered molecules is used. This direct verification of infall motion will be subject to a forthcoming publication.

References

1. Abel, T., Wandelt, B.D., 2002, MNRAS 330, L53
2. Baes, M., Stamatellos, D., Davies, J.I., Whitworth, A.P., Sabatini, S., Roberts, S., Linder, S.M., Evans, R., 2005, New Astronomy 10, 523
3. Bethell, T.J., Chepurnov, A., Lazarian, A., Kim, J., 2007, ApJ 663, 1055
4. Beuther, H., Steinacker, J., 2007, ApJL 656, L85
5. Bianchi, S. 2007, A&A 471, 765
6. Bonnell, I.A., Bate, M.R., 2002, MNRAS 336, 659
7. Chini, R., Hoffmeister, V., Kimeswenger, S., Nielbock, M., Nürnberger, D., Schmidtobreick, L., Sterzik, M., 2004, Nature 429, 155
8. Chini, R., Hoffmeister, V.H., Nielbock, M., Scheyda, C.M., Steinacker, J., Siebenmorgen, R., Nürnberger, D., 2006, ApJL 645, L61
9. Draine, B.T., Lee, H.M., 1984, ApJ 285, 89
10. Dullemond, C.P., Turolla, R., 2000, A&A 360, 1187
11. Efstathiou, A., Rowan-Robinson, M., 1995, MNRAS 273, 649
12. Ercolano, B., Barlow, M.J., Storey, P.J., 2005, MNRAS 362, 1038
13. Gordon, K.D., Misselt, K.A., Witt, A.N., Clayton, G.C., 2001, ApJ 551, 269
14. Ivezic, Z., Groenewegen, M.A.T., Men'shchikov, A., Szczerba, R., 1997, MNRAS 291, 121
15. Jonsson, P., 2006, MNRAS 372, 2
16. Juvela, M. 2005, A&A 440, 531
17. Min, M., Hovenier, J.W., de Koter, A., 2003, A&A 404, 35
18. Misselt, K.A., Gordon, K.D., Clayton, G.C., Wolff, M.J., 2001, ApJ 551, 277
19. Nielbock, M., Chini, R., Hoffmeister, V.H., Scheyda, C.M., Steinacker, J., Nürnberger, D., Siebenmorgen, R., 2007, ApJL 656, L81
20. Pascucci, I., Wolf, S., Steinacker, J., D ullemond, C.P., Henning, T., Niccolini, G., Woitke, P., Lopez, B. 2004, A&A 417, 793
21. Pinte, C., Ménard, F., Duchêne, G., Bastien, P., 2006, A&A 459, 797
22. Stamatellos, D., Whitworth, A.P., Bisbas, T., Goodwin, S., 2007, A&A 475, 37
23. Steinacker, J., Thamm, E., Maier, U., 1996, JQSRT 97, 56
24. Steinacker, J., Hackert, R., Steinacker, A., Bacmann, A., 2002, JQSRT 73, 557
25. Steinacker, J., Michel, B., Bacmann, A., 2002, JQSRT 74, 183
26. Steinacker, J., Bacmann, A., Henning, T. 2002, JQSRT 75, 765
27. Steinacker, J., Henning, T., Bacmann, A., Semenov, D., 2003, A&A 401, 405
28. Steinacker, J., Lang, B., Burkert, A., Bacmann, A., Henning, T., 2004, ApJL 615, L157
29. Steinacker, J., Bacmann, A., Henning, T., Klessen, R., Stickel, M., 2005, A&A 434, 167
30. Steinacker, J., Bacmann, A., Henning, T., 2006, ApJ 645, 920
31. Steinacker, J., Chini, R., Nielbock, M., Nürnberger, D., Hoffmeister, V., Huré, J.-M., Semenov, D., 2006, A&A 456, 1013
32. Whitney, B.A., Wood, K., Bjorkman, J.E., Wolff, M. J., 2003, ApJ 591, 1049
33. Wolf, S. 2003, Comput Phys Commun 150, 99
34. Wolfire, M.G., Cassinelli, J. P., 1987, ApJ 319, 850
35. Wood, K., Mathis, J.S., Ercolano, B. 2004, MNRAS 348, 1337
36. Yorke, H.W., Sonnhalter, C., 2002, ApJ 569, 846

Large-Scale Jet Simulations

Turlough P. Downes

Abstract In this lecture we look at the progress that has been made in studying jets with the aid of large-scale simulations. With modern computational resources and techniques it is now possible to simulate jets with reasonable resolution to a realistic age. This opens the possibility of finding out what effects a stellar jet has on its surroundings, as well as studying the stability of these jets even taking account of sophisticated chemistry occurring in the flows.

1 Introduction

In this chapter we will look at the motivations for, and results of, using large-scale numerical simulations to study the nature of jets. The phrase "large-scale simulations" can mean various things: simulations of large (or old) jets, high-resolution simulations of young jets or even moderate resolution simulations incorporating complex chemistry of young jets. For our purposes we will take the first meaning and assume that "large-scale" refers to both the simulation *and* the jet.

1.1 Motivation for Studying Large Jets

In the early days of the study of stellar jets it was believed that jets were generally only fractions of a parsec long, ranging from 0.01 pc to a few tenths of a parsec with a typical value of 0.1 pc generally being accepted [37, 35]. Even in these early studies, however, there was some hint that jets could be longer with a measured length of 2 pc for Z CMa [41].

It was well recognised, though, that the measured lengths of jets were a lower limit – the main hint for this being the observations of episodic outflows such as HH212 or HH34, first found to be a parsec-scale outflow by [2]. In these

T.P. Downes (✉)
School of Mathematical Sciences & National Centre for Plasma Science & Technology, Dublin City University, Glasnevin, Dublin 9, Ireland, turlough.downes@dcu.ie

Downes, T.P.: *Large-Scale Jet Simulations*. Lect. Notes Phys. **791**, 137–153 (2009)
DOI 10.1007/978-3-642-03370-4_5

observations it was clear that an episodic outflow can have several bowshocks associated with it – each arising from the interaction of one episode of outflow with the material in front of it which need not necessarily be ambient material. So it was clear that observing a bowshock did not mean that the end of the jet had been found. The limitations of the observing instruments, both in terms of their sensitivity and the size of the detectors made observations of large-scale flows rather difficult. The former made detection of bowshocks far from the source very difficult as their emission will be weak both due to loss of kinetic energy as they travel and also due to the tenuous nature of the medium into which they are propagating, while the latter made it difficult to "join the dots" in the overall pattern of a very large flow.

However, with the advent of more sensitive, larger instruments the number of observed parsec-scale jets has grown dramatically (e.g. [3, 16, 46, 33]) and it is not unusual now for jets to have measured lengths close to 10 pc (including both the jet and counter-jet).

It is clear, then, that very large jets exist and are not that unusual. There are several questions which arise immediately from this fact which can only be answered by large-scale numerical simulations of these systems.

The structure of this lecture is as follows: in Sect. 2 we discuss a few questions we might like answered, Sect. 3 discusses how we can ask these questions, Sect. 4 contains a discussion of some computational approaches commonly used, Sect. 5 outlines the physical conditions prevailing in jet systems and Sects. 6, 7, 8 discuss what we know so far from numerical simulations in relation to three of these questions.

2 What Questions Can We Ask?

There are an enormous number of possibilities when it comes to choosing interesting questions to investigate arising from the observations of large-scale jets. These range from the implications for accretion disk evolution through to the stability of the jets themselves. For the purposes of this discussion we will restrict ourselves to just three.

2.1 Jet Stability

One of the most striking questions relates simply to the length of the jets – can such long jets actually exist? This might seem like a rather academic point. After all, we see them so it would appear that it must be possible for them to exist. However, we must be sure that our ideas of what jets are remain consistent with the observations. So it is important to ask whether supersonic, collimated flows of gas are stable enough to propagate in a coherent way for such large distances.

We can partly address this question using the analytic techniques associated with linear stability analysis (e.g. [21]). However, given the lengths of these jets, the timescales for which they propagate and the strongly nonlinear nature of the system of equations governing them to get a full answer to this question we will need to use computational techniques.

2.2 Jet-Driven Molecular Outflows

Another issue arises when considering these jets: how do they interact with their environment? This interaction is interesting for various reasons. One is that it is well known that jets and their less well-collimated cousins molecular outflows are, observationally, closely related. There is a high coincidence of the two. In contrast to jets, molecular outflows are relatively poorly collimated (with length to width ratios of less than 5 or so), rather cold (having typical temperatures of $10 - 100$ K) and slow (with velocities of around $10 - 50$ km s^{-1}). The measured *dynamical* ages of molecular outflows range from $\sim 10^3$ to 10^5 years and their size ranges from as little as a few hundredths of a parsec to several parsecs (e.g. [32, 20, 48]).

Since jets and molecular outflows are often observed in the same places in space it is reasonable to ask whether jets could possibly drive the molecular outflows through, for example, so-called "direct" entrainment whereby the bowshock of the jet is identified as the molecular outflow (e.g. [31, 6, 42]). Given the size and age of the outflows and, by implication, the jets which are postulated to drive them large-scale simulations are certainly necessary to study this topic in depth.

2.3 Turbulence in Molecular Clouds

Turbulence is believed to be an important factor in determining the evolution of molecular clouds, both in terms of their global behaviour (such as collapse) and also in terms of the initial mass function of the stars formed in such clouds. There are various possible sources for this turbulence (see, e.g. [1, 54]), but given that the estimated energy and momentum contained in jets from young stellar objects are probably sufficient to drive the turbulence in molecular clouds (e.g. [3]) it is a fascinating question as to whether this is actually how turbulence in molecular clouds may be maintained. If it is then we have a direct feedback mechanism leading to self-regulated star formation.

Again, since the interaction between a stellar jet and its environment is such a complicated one, and since it takes place over such long time-scales, we will need to employ large-scale simulations to study it.

3 How Do We Ask Those Questions?

Once we have decided what questions we are going to ask we need to figure out how we are going to address them. In the questions that we posed in Sect. 2 there are a few commonly used ways of proceeding.

The first, and simplest, is to use hydrodynamic models. In these models we assume that **B** fields are unimportant for our purposes (or, equivalently, that since they are so poorly known they are essentially a free parameter which is chosen to be zero!). We typically assume inviscid flow (almost always a good approximation for flows in the ISM) and usually solve some kind of a chemical network in tandem with the hydrodynamic equations in order to be able to calculate emission characteristics and the radiative cooling to some reasonable degree of accuracy. We can then use the fact that we only have to solve the Euler equations in order to investigate the system of interest over longer timescales and/or with higher resolution.

So we solve the following system of equations:

$$\frac{\partial \rho}{\partial t} = -\nabla \cdot (\rho \mathbf{u}) \tag{1}$$

$$\frac{\partial (\rho \mathbf{u})}{\partial t} = -\nabla \cdot (\rho \mathbf{uu} + P\mathbf{I}) \tag{2}$$

$$\frac{\partial e}{\partial t} = -\nabla \cdot [(e + P)\mathbf{u}] - L \tag{3}$$

which are the usual inviscid Euler equations with ρ, $\mathbf{u}$, P, e and $\mathbf{I}$ being the mass density, velocity, pressure, total energy density and identity matrix, respectively. The function L represents the energy losses and gains due to other processes such as radiation, ionisation/recombination, molecular dissociation and so on. In general it will be a function of density, temperature and number densities of the various species present in the fluid and so on. In addition conservation equations for these species will usually be solved in conjunction with the Euler equations.

The system can be simplified greatly if the densities being simulated are sufficiently high that the cooling length is much less than any physical length scale of interest, but not so high that radiative transfer effects are important. In this case we can assume that the gas is isothermal and this means that we do not need to deal with Equation (3) at all. However, if we make this approximation we can no longer track any chemistry or emission properly. We also need to keep in mind that the assumption of isothermality will qualitatively affect the nature of any turbulence we may choose to investigate [17].

If we decide that hydrodynamic models do not encompass the physics we feel is important in the system of interest, then we can include the effect of magnetic fields. The usual way of doing this is to assume zero resistivity (i.e. that the magnetic field and fluid are perfectly coupled to each other). This, along with some other reasonable assumptions, leads to the ideal magnetohydrodynamic equations (see, e.g. [39] for a detailed discussion of the derivation):

$$\frac{\partial \rho}{\partial t} = -\nabla \cdot (\rho \mathbf{u}) \tag{4}$$

$$\frac{\partial (\rho \mathbf{u})}{\partial t} = -\nabla \cdot [\rho \mathbf{uu} + P^*\mathbf{I} - \mathbf{BB}] \tag{5}$$

$$\frac{\partial e}{\partial t} = -\nabla \cdot \left[(e + P^*)\mathbf{u} - (\mathbf{u} \cdot \mathbf{B})\mathbf{B} \right] - L \tag{6}$$

$$\frac{\partial \mathbf{B}}{\partial t} = -\nabla \cdot (\mathbf{uB} - \mathbf{Bu}) \tag{7}$$

$$\nabla \cdot \mathbf{B} = 0 \tag{8}$$

where all symbols are as defined in Equations (1), (2), (3), **B** is the magnetic field and

$$P^* = P + \frac{1}{2}B^2 \tag{9}$$

$$e = \frac{1}{2}\left(\rho u^2 + B^2\right) + \frac{P}{\gamma - 1} \tag{10}$$

Solving these equations is much more computationally expensive than solving Equations (1), (2), (3), and there are specific difficulties with maintaining satisfaction of the constraint Equation (8).

There are various extensions of these MHD equations to cases of non-zero resistivity (see, e.g. [7, 18, 38]) but we will not discuss these here.

4 Computational Approaches

The computational approach used is partly determined by the set of equations we decide to solve. However, for the purposes of jet simulations, there are some general guidelines.

It is generally believed to be a good idea to use second- or higher-order conservative, shock-capturing schemes (e.g. [53, 8]). The reason for using higher-order schemes is that these methods supply high accuracy, usually with less cost than using a lower order scheme with higher resolution. The emphasis on using conservation arises from the fact that by ensuring conservation of the relevant quantities (mass, momentum and energy) we ensure that the shocks will travel at the right speed and have the right strengths. This is because the Rankine–Hugoniot conditions, which are the defining relations for any shock, are derived solely from conservation. If we do not have a conservative scheme then we are not guaranteed that the shocks in our simulations will travel at the right speed and have the right post-shock conditions. There is also the rather worrying fact that some non-conservative schemes in use in popular codes today are actually not consistent in the sense that the numerical error in the simulation does not tend to zero as we increase our spatial and temporal resolution [23]).

It could be argued that since, in the case of our simulations, we do not have a conservative system anyway (energy is not conserved due to the term denoted by L in Equations (3) and (6)) the enthusiasm for conservation may not be so well justified. However, no such argument rescues us from the latter problem. Note that not all codes commonly in use today use conservative schemes (e.g. [51]).

The most common modern algorithms for solving the hydrodynamic or MHD equations today are very computationally intensive, involving the solution of Riemann problems at each zone interface in the simulation. Since the cooling lengths in jet simulations are frequently much less than the jet radius and since the length-to-width ratio of the jets themselves is so large, we need very high resolution. This translates into needing an enormous amount of memory and CPU cycles to do useful simulations. In fact, there is no way to perform simulations of the systems suggested in Sect. 2 to reasonable accuracy without using some clever computational techniques to minimise the wall-clock time to solution and/or the memory necessary.

The two most common ways of reducing the computational effort involved are parallelisation and/or adaptive mesh refinement. Parallelisation can be achieved in a relatively straightforward fashion for the (magneto)hydrodynamic equations as they are strictly hyperbolic. Hence, once we use explicit numerical methods, only local interactions are important. The parallelisation can then be done efficiently by domain decomposition in which the computational domain is divided into equal-sized blocks and distributed out to the processors performing the simulation. Each then integrates the equations on its own part of the domain, swapping only boundary data with its neighbouring processes as necessary (Fig. 1).

Note, however, that if non-local effects (such as radiative transfer or self-gravity) are included then processors will need to swap much larger amounts of data as the simulation proceeds, increasing the communication overhead for the simulation and therefore its time-to-completion. Similarly, implicit numerical methods are quite complicated to implement efficiently and accurately in a parallel environment.

Adaptive mesh refinement (AMR) adopts a different approach. The idea is to find "interesting" regions in the simulation (such as shocks, regions of high radiative cooling and steep density gradients) and add new computational zones to those areas. These zones are then deleted once the interesting feature has moved to a different part of the grid. This approach can reduce the number of zones needed for a given simulation quite significantly, depending on the nature of the simulation being performed. This leads to a reduction in the number of calculations and also in the memory necessary.

An obvious thing to do is to implement both parallelisation and AMR. This is quite tricky to do successfully due to the uneven spread of grid points in space making load-balancing difficult.

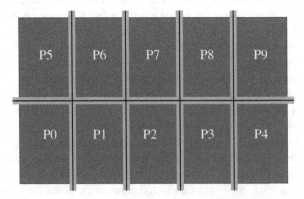

Fig. 1 (Color online) Illustration of domain decomposition. Since the systems of equations in question are hyperbolic and assuming explicit schemes are used, it is only necessary for processors to communicate their boundaries with neighbouring processors (the *dark-grey shaded strips*)

One way around using AMR with parallelisation and its attendant challenges is valid in the case of jets. Since we know that jets initially start off small and get larger and longer we can use that information to construct our grid: we use a grid only just large enough to contain the jet/bowshock structure at any given time. We then distribute this grid among the available processors (see Fig. 2). Since the jet expands roughly linearly in time in both its longitudinal and transverse directions this will give a time-saving of approximately 4 when running simulations in two dimensional and 8 when running three dimensional simulations.

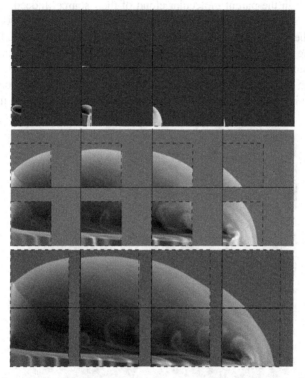

Fig. 2 Illustration of domain decomposition in the case where we ensure that calculations only occur in a grid just large enough to contain the jet/bowshock system. The age of the simulated jet increases from the *top panel* to the *bottom panel* and the *dashed lines* indicate the the domain actually being computed by each process

5 Physical Conditions for Jet Simulations

We now turn to the issue of the physical conditions thought to be relevant for jet simulations. The estimates of these conditions, in particular the densities, have varied quite a bit over the last 20 years. The relevant conditions also depend on what kind of jet we want to simulate – very young jets can be denser and possibly slower than their more evolved counterparts.

We split our discussion of these conditions into two: those relevant for the ambient medium and those relevant for the jet material itself.

5.1 Ambient Conditions

The density of the ambient medium varies a lot. Close to the star the densities can be quite high ($\geq 10^4$–10^5 cm^{-3}). This corresponds to the molecular cloud surrounding the protostar. However, as can be seen from Fig. 3 jets from young stars do not always remain in the parent molecular cloud of their source. Indeed, we certainly would not expect them to given that we are simulating quite long jets which will have exited the molecular cloud and begun propagating into the interstellar medium external to the cloud. The relevant densities for this kind of a region can be taken as typically around 1 cm^{-3}.

Along with the variation in density there will be a variation in the chemical makeup of the ambient material into which the jet is propagating. Close to the protostar

Fig. 3 (Color online) Observation of jets in the Carina nebula. It is clear that these jets have exited the molecular cloud and are now propagating into a HII region which is produced by a nearby O star. Credit: Nathan Smith and John Bally

and assuming there are not any strong radiation fields, the material will be almost completely molecular, weakly ionised and will contain dust grains. Once the jet has broken out of the parent molecular cloud, though, the ambient material will be almost completely atomic.

The temperature will also vary from around $10-20$ K close to the star to around 10^4 K outside the cloud. This also means that the sound speed will change from only a few hundred metres per second up to around 10 km s^{-1}.

Finally, the magnetic field will also vary. It is not really known how it will vary though, and neither is its topology well constrained by observations or theory on the large scales of interest to us in these simulations. However, it is reasonable to say that outside the molecular cloud we would typically expect the magnetic field strength to be of the order of μG and that it is likely to be significantly higher than this close to the source of the outflow.

5.2 Jet Conditions

The densities of jets, assuming an average ionisation fraction of 0.1, are thought to be in the region of 10^3–10^4 cm^{-3} [40]. The temperatures are generally accepted to be in the range 10^3–10^4 K. These ranges of densities and temperatures imply that radiative cooling will be dynamically significant. Many researchers have included complex chemical reaction networks in their codes in order to allow for proper treatment of this effect (e.g. [5, 11, 34]).

Jet velocities are reasonably well constrained and generally lie in the range $100-400$ km s^{-1} (e.g. [37, 35]). It is by now well accepted that the blobs of emission along the lengths of jets are caused by variations in the jet velocity (e.g. [43]). These variations can range from a few km s^{-1} up to virtually the jet velocity itself, resulting in episodic outflows along the jet. This episodicity appears to be quite a general phenomenon, at least in observations of parsec-scale jets (e.g. [2, 33]) and occurs on quite long time-scales. Of course, one would not expect to see such a large amount of evidence for episodicity in observations of shorter jets given the long time-scale associated with any observed episodicity.

In addition to variations in the magnitude of the jet velocities, variations in direction are frequently observed (see Fig. 4). Simulation of these systems requires a fully three dimensional code and is therefore quite computationally demanding (see, e.g., [29]).

The magnetic field in the jet material itself is not well-constrained. Observations in [44] suggest that field strengths will be of the order of a few Gauss. Assuming that jets are magnetocentrifugally launched one would expect that the fields will be largely toroidal in nature within the jet itself. However, there are no observations which directly confirm this hypothesis. Assuming we can use ideal magnetohydrodynamics, the field must be dynamically significant within the jet. After all, it is precisely this field which caused the jet to collimate close to the star and which is now being advected out along the jet with the jet material itself.

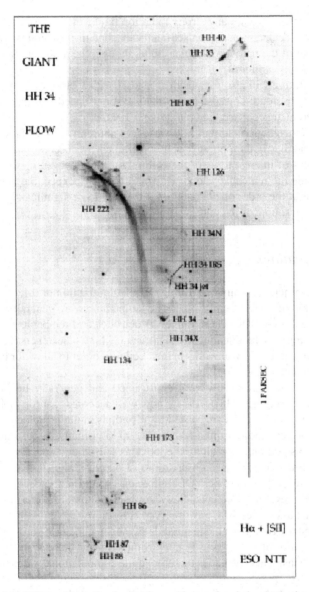

Fig. 4 The giant HH34 outflow, clearly showing evidence of variation in the jet direction with time, resulting in the S-shaped morphology. Adopted from Devine et al. [11]

There is some question over whether, for parsec-scale jets, ideal magnetohydro-dynamics is appropriate [19]. The timescale for ambipolar diffusion in jet material is about

$$t_{\mathrm{ad}} \approx 3 \times 10^4 \left(\frac{n}{10^3 \mathrm{cm}^{-3}} \right) \left(\frac{R_{\mathrm{jet}}}{10^{15} \mathrm{cm}} \right) \left(\frac{10^4 \mathrm{K}}{T_{\mathrm{jet}}} \right) \left(\frac{\beta}{\beta + 1} \right) \ \mathrm{years} \qquad (11)$$

which is of the same order as the length of time it will take material (and hence the magnetic field itself) to be advected along the full length of a parsec-scale jet. Hence ambipolar may play some role in the behaviour of the system, albeit almost certainly a fairly minor one.

Finally, it is worth noting that, were weak magnetic fields to be detected in stellar jets at large distances from the source this would not mean that these fields are not significant. This is due to the apparent scaling of the magnetic field with n^p, where n is the number density and $0.5 < p < 1$. Hence rarefactions along the jet's length, which almost certainly occur, may lead to observations of weak fields where, in fact, they are dynamically important [22].

Now that we have some idea of the kinds of initial conditions to use we will go on to discuss the specific problems outlined at the start of this lecture.

6 Stability of Jets

We now turn to the question of whether or not parsec-scale jets "should" exist from the point of view of their stability. It is not immediately obvious that a long, narrow, fast moving column of gas would succeed in propagating for large distances given the wide variety of fluid instabilities which exist.

However, one thing which can reassure us, at least a priori, is that we know that supersonic jets are self-collimating – if a supersonic jet begins to decollimate shocks are driven into it which tend to force it to recollimate again. An example of this can be seen in Fig. 5 which shows a picture of a NASA SR71 in flight. The exhaust from the engines is supersonic (with respect to the atmosphere). Initially, as it exits the engines, the exhaust is overpressured. As it expands it becomes under-pressured with respect to the atmosphere and a conical shock is driven inward. This raises the

Fig. 5 (Color online) A photograph of a NASA SR71 in flight. Note the blobs of emission along the length of the engine exhausts, called Mach diamonds. See text

pressure in the exhaust and the process repeats. The shocks crossing the exhaust are clearly visible as the blobs of yellow emission along its length. These are also known as Mach diamonds.

Given that we expect a large shear to occur at the walls of a jet the most obvious candidate for an instability to decollimate the jet is the Kelvin–Helmholtz (KH) instability. This is an instability which may occur whenever there is shear in a flow. Much work has been done on this both analytically and numerically (e.g. [30, 47, 14, 52]). The basic results which can be taken from this work are as follows:

- the KH instability grows strongly only for relatively low Mach number jets,
- it grows strongly only for flows where the density ratio between the jet and the ambient medium is close to one, and
- radiative cooling stabilises jets against the instability [21, 47].

Contrast these requirements with the conditions for YSO jets:

- YSO jets have Mach numbers in the range ~ 20–100,
- the density ratio between a YSO jet and the material surrounding it (i.e. the cocoon of the jet) is very high (even if the density ratio between the jet and the ambient material is of order one), and
- radiative cooling is dynamically significant in jets.

We can see that the KH instability is unlikely to be effective in decollimating YSO jets. Indeed long-duration simulations of YSO jets do not show any sign of the KH instability (see, for example, Fig. 6).

Fig. 6 An axisymmetric simulation of a YSO jet at age 2300 years. There is no sign of the growth of any instability likely to disrupt the jet (taken from [24])

So we can conclude that, on the basis of both analytic work and on both small and large-scale simulations of YSO jets, even on very long time-scales YSO jets do not need any stabilising mechanism to maintain their collimation. Hence the existence of parsec-scale jets do not pose any challenges to theory, at least in terms of their stability.

7 Jet-Driven Molecular Outflows

We now turn to the issue of molecular outflows. Various models have been proposed to explain why molecular outflows are so closely associated with YSO jets. One of the first models suggested was the so-called "steady-state entrainment" where the molecular outflow would be ambient material accelerated at the boundaries of

YSO jets through the growth of instabilities. As we have seen in Sect. 6 this is unlikely to work well, given how stable YSO jets appear to be. Also, in order to accelerate a massive molecular outflow they would have to lose approximately half of their momentum through instabilities without actually losing their collimation – a seemingly unlikely scenario.

One of the most popular current models, particularly for relatively well-collimated molecular outflows, is that the molecular outflow is actually the bowshock of a YSO jet (e.g. [42]). The idea is that the jet drives a bowshock into the molecular cloud as it propagates. This bowshock consists of ambient (i.e. molecular cloud) material. The material in the bowshock is itself the molecular outflow.

The questions we can ask regarding this model are the following:

- Is the overall morphology of the "molecular outflow" produced in this model similar to those observed?
- Are the line profiles and position–velocity diagrams one gets from such a molecular outflow model the same as those observed?

Many authors have performed simulations intended to investigate these questions (e.g. [6, 50, 15, 26, 12, 49, 36]), but usually for timescales of $\leq 10^3$ years. Generally speaking the results of these simulations match the observations remarkably well, although the precise slopes of the wings of the line profiles was noted to be time-dependent.

Longer duration simulations (up to 2300 years with a resolution of 10^{14} cm using the technique outlined at the end of Sect. 5.1) were subsequently performed by [24] who found that these slopes settled to a fairly constant value after around 1500 years. However, it was noted from the simulations (see Fig. 7) that the length-to-width ratio of the molecular outflows produced in this way was rather large (see also [26, 49]). This issue can, however, be significantly ameliorated by invoking an episodic jet as the driving mechanism for the outflow.

Further large-scale simulations were performed by [13] who noted that the ages, at least of molecular outflows driven by jets still contained within their parent molecular cloud, may be overestimated by as much as an order of magnitude by standard

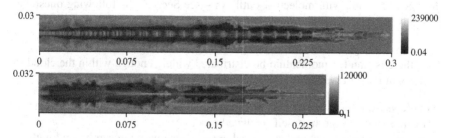

Fig. 7 Simulations of the jet-driven bowshock model for relatively long timescales. Note that the length-to-width ratio of the bowshock (i.e. the molecular outflow) is rather large, although it can be reduced if the jet is taken to be episodic (*bottom panel*). Taken from [24]

methods. Despite this, the simulations show that the momentum budget is likely to be about right for stellar jets to be able to drive these outflows.

Another possible mechanism for generating molecular outflows, closely related to the jet-driven model, is that involving a wide-angled wind. In this model a wide-angled wind plays the role of the jet. One strong benefit of this model is the ability to produce molecular outflows with lower length-to-width ratios more appropriate for some observed outflows. See [26] for a nice comparison of the jet- and wind-driven models.

It is probably fair to say that, while it is generally accepted that molecular outflows are driven by outflows from YSOs, the details of whether the driving outflows should be wide-angled winds or well-collimated jets are not yet known.

8 Turbulence Generation

Turbulence in molecular clouds is a much-studied subject – so much so that it is beyond the scope of this work to give even a reasonably complete overview of the current state of the art in this area. One of the reasons for the interest in this topic is that turbulence may act to support a molecular cloud against gravitational collapse, hence effecting the formation of stars. The reader is referred to [17] for an extensive review of the study of astrophysical turbulence.

It is worth noting at this point that not everyone believes that the motions of the gas and dust that we observe in molecular clouds is, formally speaking, turbulence. However, there are some observations which lead one to the belief that it is at least a reasonable starting point to presume that what we observe in molecular clouds is actually turbulence. One of the most compelling of these is the phenomenon that velocity dispersion in observations of emission from molecular clouds increases with increasing field size as a power-law similar to what would be expected from compressible MHD turbulence [25].

We know from simulations (e.g. [28, 27]) that supersonic turbulence decays quickly. As a result, if turbulence is present in molecular clouds then the energy and momentum in it must be replenished by some process. Since jets from young stars can have a significant impact on the structure of molecular clouds (e.g. with their probable link with molecular outflows – see Sect. 7) the following questions naturally arise:

- Is there enough momentum in jets to drive cloud turbulence?
- If there is, can the momentum be distributed widely enough within the cloud to account for observations?

If outflows from forming stars can drive the turbulence in molecular clouds then star formation may be a strongly self-regulated process.

In terms of theoretical and numerical approaches one thing can be said without fear of contradiction – it is not easy to make significant progress in understanding turbulence:

- It is very difficult to study analytically:

  - Statistical approaches do not have satisfactory closure relations.
  - Even partially successful approaches tend to fail for supersonic turbulence.
  - The situation is much more complicated for MHD flows.
  - It is even less tractable for compressible MHD flows.

- It cannot be accurately simulated:

  - The dynamic range involved in turbulence is truly enormous.
  - Simplifying assumptions (e.g. using an isothermal approach – see Sect. 3) have significant qualitative effects on the results.
  - Periodic boundary conditions, frequently used in turbulence simulations, also qualitatively effect the results.

These difficulties coupled with the likely astrophysical significance of turbulence do, however, make it a fascinating subject to study.

It appears that jets from YSOs have the energy to drive supersonic turbulence in molecular clouds. That is, if one examines the rate of decay of this turbulence and compares it with the injection rate of momentum into the molecular cloud by jets then we arrive at a rough equality (e.g. [45]). So the question arises that if jets could in principle be an important source of turbulence in molecular clouds then how might it work in practice? [27]

Using resolutions of 128^3, concluded that turbulence driven by collimated outflows decays more rapidly than that driven on a large scale. These simulations, as noted by the author, are of low resolution – the size of the "jets" being approximately five computational zones. Hence we must treat this result with some caution.

Subsequently [9] performed simulations using 3000×650 zones to examine cavities left in molecular clouds after jets have ceased (so-called "fossil cavities"). The aim was to investigate how turbulence may propagate from the site of the jet into the cloud. These authors found that the resulting turbulence was mainly subsonic in nature. Given the supersonic nature of the observations of motions in molecular clouds this presents a considerable difficulty.

Banerjee et al. [4] performed quite high resolution simulations using AMR (equivalent to 3072×1024 grid zones) of single jets propagating into an ambient medium and found that the resulting turbulence appeared to be small-scale in the sense that it remained fairly local to the jet/bowshock system. This result is consistent with the results of [9] and adds to the difficulties of the notion that stellar jets can drive the observed turbulence in molecular clouds.

To summarise, then, as things stand at the moment it would appear unlikely that jets from YSOs can drive the large scale turbulence observed in molecular clouds. However, given all the problems associated with simulating turbulence and, in particular, the relatively low resolutions being used, it would be wrong to view this as the final answer to the issue of outflow-driven turbulence.

9 Conclusion

We can conclude our discussion of large-scale simulations of jets as follows:

- With many modern problems in jets and star formation it is necessary to have both high resolution and a physically large domain. These two requirements make parallel and/or AMR methods highly desirable for the study of these problems.
- Results of the simulations of large-scale jets show that hydrodynamic jets do seem to be stable enough to propagate for several parsecs without breaking up – they do not need any additional collimation mechanisms to maintain their integrity. These numerical results are consistent with indications from fluid dynamics theory, although it is true that there are still some unanswered questions in this area.
- It would appear that jets are capable of driving many molecular outflows. There are some problems arising from the fact that jets appear to produce rather narrow outflows but it is possible that this issue can be ameliorated if the outflow is being driven by an episodic jet.
- The question of whether YSO jets can drive molecular cloud turbulence is still an open one – indeed, there is still some debate over whether the motions observed in molecular outflows are truly turbulent. Further work is needed in this area, in particular with higher-resolution simulations. The results obtained to date indicate that it is rather difficult to obtain large scale turbulence of the type observed using outflows from YSOs.

Acknowledgements This work was partly funded by the CosmoGrid project, funded under the Programme for Research in Third Level Institutions (PRTLI) administered by the Irish Higher Education Authority under the National Development Plan and with partial support from the European Regional Development Fund. The present work was also supported in part by the European Community's Marie Curie Actions – Human Resource and Mobility within the JETSET (Jet Simulations, Experiments and Theory) network under contract MRTN-CT-2004 005592.

References

1. Ballesteros-Paredes, J., Klessen, R.S., Mac Low, M.-M., Vazquez-Semadeni, E., 2007, Molecular Cloud Turbulence and Star Formation. In: Protostars and Planets V, B. Reipurth, D. Jewitt, and K. Keil (eds.), University of Arizona Press, Tucson, 63–80
2. Bally, J., Devine, D., 1994, ApJ 428, L65–L68
3. Bally, J., Devine, D., Alten, V., 1996, ApJ 473, 921
4. Banerjee, R., Klessen, R.S., Fendt, C., 2007, ApJ 668, 1028–1041
5. Biro, S., Raga, A.C., Cant'o, J., 1995, MNRAS 275, 557–566
6. Chernin, L., Masson, C., Gouveia dal Pino, E.M., Benz, W., 2004, ApJ 426, 204–214
7. Ciolek, G.E., Roberge, W.G., 2002, ApJ 567, 947–961
8. Colella, P., Woodward, P.R., 1984, J Comp Phys 54, 174–201
9. Cunningham, A.J., Frank, A., Quillen, A.C., Blackman, E.G., 2006, ApJ 653, 416–424
10. De Colle, F., Raga, A.C., 2006, A&A 449, 1061–1066
11. Devine, D., Bally, J., Reipurth, B., Heathcote, S., 1997, AJ, 114, 2095
12. Downes, T.P., Cabrit, S., 2003, A&A 403, 135–140
13. Downes, T.P., Cabrit, S., 2007, A&A 471, 873–884

14. Downes, T.P., Ray, T.P., 1998, A&A 331, 1130–1142
15. Downes, T.P., Ray, T.P., 1999, A&A 345, 977–985
16. Eislöffel, J., Mundt, R., 1997, AJ 280, 280–287
17. Elmegreen, B.G., Scalo, J., 2004, ARA&A 42, 211–273
18. Falle, S.A.E.G., 2003, MNRAS 344, 1210–1218
19. Frank, A., Gardiner, T.A., Delamarter, G., Lery, T., Betti, R., 1999, ApJ 524, 947–951
20. Gueth, F., Guilloteau, S., 1999, A&A 343, 571–584
21. Hardee, P., Stone, J.M., 1997, ApJ 483, 121–135
22. Hartigan, P., Frank, A., Varniére, P., Blackman, E.G., 2007, ApJ 661, 910–918
23. Hou, T.Y., LeFloch, P.G., 1994, Math Comput 62, 497–530
24. Keegan, R., Downes, T.P., 2005, A&A 437, 517–524
25. Larson, R.B., 1981, MNRAS 194, 809–826
26. Lee, C-F, Stone, J.M., Ostriker, E.C., Mundy, L.G., 2001, ApJ 527, 429–442
27. Mac Low, M.-M., 2000, Turbulence Driven by Stellar Outflows. In: F. Favata, A. Kaas, and A. Wilson (eds.) Proceedings of the 33rd ESLAB symposium on star formation from the small to the large scale, ESTEC, Noordwijk, The Netherlands, 2–5 November 1999 Noordwijk, The Netherlands: European Space Agency (ESA), p. 457–460
28. Mac Low, M.-M., Klessen, R.S., Burkert, A., Smith, M.D., 1998, Phys. Rev. Lett. 80 2754–2757
29. Masciadri, E., de Gouveia Dal Pino, E.M., Raga, A.C., Noriega-Crespo, A., 2002, ApJ 580, 950–958
30. Massaglia, S., Trussoni, E., Bodo, G., Rossi, P., Ferrari, A., 1992, A&A 260, 243–249
31. Masson, C.R., Chernin, L.M., 1993, ApJ 414, 230–241
32. McCaughrean, M.J., Rayner, J.T., Zinnecker, H., 1994, ApJ 436, L189–L192
33. McGroarty, F., Ray, T.P., Bally, J., 2004, A&A 415, 189–201
34. Mignone, A., Bodo, G., Massaglia, S., Matsakos, T., Tesileanu, O., Zanni, C., Ferrari, A., 2007, ApJS 170, 228–242
35. Mundt, R., 1993, Observational Properties of Jets from Young Stars. In: L. Errico and A. Vittone (eds.) Stellar Jets and Bipolar Outflows, Proceedings of the 6th International Workshop of the Astronomical Observatory of Capodimonte (OAC 6), held at Capri, Italy, September 18–21, 1991, p. 91. Kluwer, Dordrecht
36. Moraghan, A., Smith, M.D., Rosen, A., 2006, MNRAS 371, 1448–1458
37. Mundt, R., Brugel, E.W., Bührke, T., 1987, ApJ 319, 503
38. O'Sullivan, S., Downes, T.P., 2007, MNRAS 376, 1648–1658
39. Pelletier, G.: Introduction to Magneto-Hydrodynamics. In: Ferreira, J., Dougados, C., Whelan, E. (eds.): Jets from Young Stars. Lect. Notes Phys. 723, 77–101 (2007)
40. Podio, L., Bacciotti, F., Nisini, B., Eislöffel, J., Massi, F., Giannini, T., Ray, T.P., 2006, A&A 456, 189–204
41. Poetzel, R., Mundt, R., Ray, T.P., 1989, A&A 244, L13–L16
42. Raga, A.C., Cabrit, S., 1993, A&A 278, 267–278
43. Raga, A.C., Cantó, J., Binette, L., Calvet, N., 1990, ApJ 364, 601–610
44. Ray, T.P., Muxlow, T.W.B., Axon, D.J., Brown, A., Corcoran, D., Dyson, J., Mundt, R., 1997, Nature 385, 415–417
45. Reipurth, B., Bally, J., 2001, ARA&A 39, 403–455
46. Reipurth, B., Bally, J., Devine, D., 1997, AJ 114, 2708–2735
47. Rossi, P., Bodo, G., Massaglia, S., Ferrari, A., 1997, A&A 321, 672–684
48. Shepherd, D., Watson, A.M., Sargent, A.I., Churchwell, E., 1998, ApJ 507, 861–873
49. Smith, M.D., Rosen, A., 2005, MNRAS 357, 579–589
50. Smith, M.D., Suttner, G., Yorke, H.W., 1997, A&A 323, 223–230
51. Stone, J.M., Norman, M.L., 1993, ApJ 413, 198–220
52. Stone, J.M., Hardee, P.E., 2000, ApJ 540, 192–210
53. van Leer, B., 1977, J Comp Phys 23, 276–299
54. Zinnecker, H., Yorke, H., 2007, ARA&A 45, 481–563

Modeling Accretion and Ejection Phenomena Around Young Stars: A Numerical Perspective

Claudio Zanni

Abstract This chapter presents an overview of models based on time-dependent numerical simulations of the central regions of accreting T Tauri stars. The focus is put on magnetohydrodynamic studies of the launching mechanism of T Tauri jets (disk–winds, stellar winds, magnetospheric ejections) and of the magnetic star–disk interaction.

1 Introduction

Collimated jets are a characteristic phenomenon often displayed by actively accreting "classical" T Tauri stars (CTTS). These outflows are observed in low excitation optical forbidden lines on scales of 10–100 AU propagating with a typical speed of the order of the escape velocity from the central star ($\sim 200\,\mathrm{km\,s^{-1}}$). Despite the clear connection with accretion [9, 32] and the richness of information about the dynamics and the thermodynamics of these sources [3, 42], the precise origin of T Tauri jets is still debated: are the jets launched from the protostar, from the surrounding accretion disk or from the region of interaction between the stellar magnetosphere and the disk?

On the other hand, all the proposed scenarios agree on one point: the presence of a large scale magnetic field is required. The field can in fact provide an effective means to extract the rotational (gravitational) energy stored in the star disk system and transfer it to the jets. Moreover, magnetohydrodynamic (MHD) processes are to date the only physical mechanisms able to provide self-confinement and collimation to the outflows without recurring to external agents: this is achieved through a hoop-stress effect, i.e., the magnetic tension of the toroidal field component arising naturally along magnetic surfaces anchored onto a rotating object.

Therefore, it is possible to discern between the proposed models by looking at the anchoring position of the magnetic surfaces and at the magnetic flux distribution.

C. Zanni (✉)

Laboratoire d'Astrophysique de Grenoble, 414 Rue de la Piscine, BP 53, F-38041 Grenoble, France, Claudio.Zanni@obs.ujf-grenoble.fr

Zanni, C.: *Modeling Accretion and Ejection Phenomena Around Young Stars: A Numerical Perspective*. Lect. Notes Phys. **791**, 155–178 (2009)

DOI 10.1007/978-3-642-03370-4_6

We talk about *disk–winds* if the plasma flows along magnetic field lines anchored onto the surface of the accretion disk: moreover, we distinguish between *extended disk–winds* [6, 72, 26] if the magnetic flux is distributed on a large radial extension of the disk, and *X-Winds* [65, 11], if the field is concentrated in a tiny region around the magnetopause. A *stellar wind* [73, 63, 50] flows along opened magnetic surfaces anchored onto the surface of the rotating star. Finally, non-stationary *magnetospheric ejections* can take place due to the inflation-reconnection-opening of the magnetospheric field lines connecting the star with the disk [30, 47].

Due to the different angular speed of the "rotator" on which the magnetic field lines are anchored, these models display different dynamical characteristics: in particular, it has been shown that, if the velocity shifts measured transversally to the jet axis are representative of jet rotation [4, 16, 74], only disk–winds launched from a radial extension 0.1–2 AU of the accretion disk are consistent with the poloidal/toroidal speed measurements of T Tauri jets [28]. Moreover, the ratio between the ejection and the mass accretion rate estimated for T Tauri jets (~0.01–0.1, [10]) is energetically less problematic if the launching region is distributed over a substantial radial extension of the accretion disk. Obviously the outflowing components emerging from the inner part of the star disk system are likely to coexist with the outer disk–winds: they can play a crucial role in explaining a high–velocity component observed in T Tauri jets [22, 20] and in regulating the angular momentum of the central protostar [49].

In this chapter I will present a general overview of the *numerical* studies available in literature to model different parts of T Tauri star–disk–jet systems.[1] Section 2 will be devoted to the problem of the launching of disk–winds, while in Sect. 3 I will present the studies dealing with the magnetic star–disk interaction and the related outflows.

2 Jet Launching from Accretion Disks

The idea of magnetohydrodynamic launching from Keplerian accretion disks was originally proposed by Blandford and Payne [6]. The authors showed how the Poynting flux along opened magnetic field lines can extract energy and angular momentum from the underlying Keplerian disk and transfer them to an outflow which is magneto-centrifugally accelerated. At the base of the outflow, where the field anchored in the disk and the plasma flowing along it co-rotate with the Keplerian disk, the centrifugal acceleration is dominant. They showed that the cen-

---

[1] All the models proposed are a solution of the ideal or resistive/viscous set of the magnetohydrodynamic (MHD) equations [58]. Recurring to a cylindrical (r, z, ϕ) system of coordinates, I will use a standard notation for the dynamical quantities: ρ indicates the matter density, $v_p = \sqrt{v_r^2 + v_z^2}$ the speed in the poloidal plane, v_ϕ the toroidal velocity; P is the thermal pressure, $B_p = \sqrt{B_r^2 + B_z^2}$ is the poloidal component of the magnetic field, while B_ϕ is the toroidal one.

trifugal thrust becomes stronger than the gravitational pull when the field lines are inclined with an angle $\theta > 30°$ with respect to the rotation axis of the disk. Close to the Alfvén critical surface, where the poloidal speed equals the Alfvén speed $V_A = B_p/\sqrt{4\pi\rho}$, the matter inertia is comparable to the magnetic energy: the matter lags behind the field rotation, a dominant toroidal component is therefore created, and the residual acceleration is due to the gradient of the toroidal field along the magnetic surfaces. The MHD torque allows the disk to accrete while the hoop-stress exerted by the toroidal field component can provide a mechanism of self-collimation. Inspired by this seminal work, a considerable theoretical effort has been done to produce more general and complete MHD models of the jet launching process. In Sect. 2.1 I will point out the general ideas concerning the analytical models, while in Sect. 2.2 I will present an overview of the numerical studies.

2.1 Jet Launching: Analytical Models

Most of the available analytical models of the MHD launching of jets from Keplerian accretion disks are based on stationary (i.e., $\partial/\partial t = 0$), axisymmetric (i.e., $\partial/\partial\phi = 0$), self-similar solutions of the MHD system of equations. Stationarity and axisymmetry imply, in an ideal MHD regime, the existence of invariants, i.e., quantities which are constant on the magnetic surfaces along which the plasma flows. These invariants are the mass-loading parameter k

$$k = \frac{4\pi\rho v_p}{B_p},\tag{1}$$

given by the mass-to-magnetic flux ratio; the rotation rate of the magnetic surfaces Ω_*, defined by the relation

$$v_\phi = r\Omega_* + v_p \frac{B_\phi}{B_p};\tag{2}$$

the specific angular momentum l:

$$l = r v_\phi - \frac{r B_\phi}{k}.\tag{3}$$

Evaluating this expression at the Alfvén surface, the specific angular momentum takes the simple form $l = r_A^2\Omega_*$, where r_A is the cylindrical radius of the Alfvén surface. Its nondimensional form is usually indicated as

$$\lambda = \frac{l}{r_0^2\Omega_*} = \left(\frac{r_A}{r_0}\right)^2,\tag{4}$$

where r_0 is the cylindrical radius of the footpoint of the magnetic surface. The ratio r_A/r_0 defines the magnetic lever arm. The specific energy (Bernoulli invariant) e is defined as

$$e = \frac{1}{2}v^2 + h + \Psi - \frac{r\Omega_* B_\phi}{k},\tag{5}$$

given by the sum of the kinetic energy, the enthalpy h, the gravitational potential energy Ψ and the work done by the magnetic field. In general these models are also polytropic, i.e. the entropy $S = P/\rho^\gamma$ is conserved along a magnetic surface.

Analytical disk-wind models assume generally a radial self-similarity of the solutions: the radial dependency of all the dynamical quantities is fixed a priori by a suitable power-law while it is necessary to solve for the vertical behavior of the solutions. Assuming that the magnetic surfaces are "frozen" in the disk Keplerian rotation, their angular speed is given by $\Omega_* = \Omega_K = \sqrt{GM/r_0^3}$. This assumption implies that all the characteristic speeds of the solution must scale radially as the Keplerian speed, i.e., $\propto r^{-1/2}$. This constraint leaves anyway a degree of freedom on the choice of the radial distribution of the magnetic flux and the density: an entire class of radially self-similar solution can therefore be constructed by varying the radial distribution of the magnetic field [17, 71]. The Blandford and Payne solution is in fact a particular case in which the thermal effects are neglected ($h = P = S = 0$) and the radial distribution of the field follows the power-law $B \propto r^{-5/4}$.

A further complication arises when we want to connect the jet solution with the physics of the underlying accretion disk. In particular the mass outflow rate, which in the launching models is controlled by the free parameter k, must be computed consistently by accurately solving the vertical equilibrium of the disk. Moreover, since we are looking for stationary solutions, diffusive effects must be introduced inside the accretion disk: these are necessary to balance the advection of the magnetic flux towards the central star with an adequate diffusion directed radially outwards. This kind of solutions have been computed by introducing diffusive effects like ambipolar diffusion [72, 43] or using an α parametrization [25, 26, 12] for the disk resistivity v_m, i.e., $v_m = \alpha_m V_A H$, where V_A is the Alfvén speed calculated at the midplane of the disk and H is the thermal heightscale of the disk. These models show clearly that the disk is pinched vertically by the gravity but also by the Lorentz force: the thermal pressure gradient is the only force which can assure the vertical quasi-equilibrium of the disk and uplift the accreting material at the disk surface where the MHD acceleration can take place. In particular the models developed by Ferreira and collaborators [25, 26, 12] have shown clearly that a stationary solution requires a magnetic field around equipartition with the thermal energy of the disk ($\mu = B^2/4\pi P \sim 1$) and a strong, even anisotropic, resistivity ($\alpha_m \sim 1$).

Analytical models have been proved to be a powerful tool to reproduce the dynamical constraints coming from the observations of T Tauri jets [28], in terms of collimation, poloidal speed and angular momentum: why therefore bother looking for numerical solutions of the problem?

2.2 Jet Launching: Numerical Studies

There is, of course, a series of questions which can be addressed only by recurring to numerical time-dependent simulations:

- Test the analytical models, mostly in the parameter range which is not allowed by stationary solutions.
- Look for non-self-similar stationary solutions, combining also different components (i.e., stellar + disk winds).
- Look for truly time-dependent solutions to address the problem of variability.
- Perform a three-dimensional study of stability of the jet-launching mechanism.

After a short note on the importance of the choice of initial and boundary conditions to perform MHD numerical simulations of the jet-launching process, I will briefly review some numerical models which have tried to address these questions.

2.2.1 The Choice of Initial and Boundary Conditions

Most of the numerical models of jet launching are axisymmetric, as the analytical solutions. Generally also the three-dimensional simulations are based on axisymmetric initial conditions including non-axisymmetric perturbations. The most popular approach to perform numerical simulations of the MHD jet launching [55, 56, 67, 39, 40, 14, 15, 75] has been to use as initial condition the rotator, i.e., a Keplerian disk, threaded by a large-scale poloidal magnetic field on top of which a non-rotating atmosphere is set. The current circuits and the toroidal field necessary to propel the jet arise because of the differential rotation between the disk and the atmosphere. The simulation is continued until it reaches a more or less stationary situation. A different and less exploited approach is to use as initial condition a solution inspired by an analytical self-similar model [31, 46]: this approach allows to test directly the validity of the stationary analytical models.

Independently of the initial setup, conditions must be set on the boundaries of the computational domain: except from conditions dictated by the symmetry of the problem, these have to be chosen carefully. The usual and correct approach requires that the number of boundary conditions which are allowed to be fixed is equal to the number of characteristic waves entering the computational domain at a specific boundary. The other quantities must be let evolve following the evolution of the solution inside the computational domain. Anyway also these free boundary conditions must be chosen carefully not to force the behavior of the numerical solution: a typical example of the effects of boundary conditions onto the characteristics of the solution is the artificial collimating effect determined by a $\partial B_\phi/\partial r = 0$ condition imposed onto the outer radial boundary of the computational domain [67, 75].

The numerical models available in literature can be divided into two groups, depending on the choice of initial and boundary conditions: simulations treating the disk rotation as a boundary condition and focusing on the launching mechanism (Sect. 2.2.2) and simulations studying the dynamics of the entire accretion–ejection process (Sect. 2.2.3).

2.2.2 MHD Simulations of the Jet Launching Mechanism

A great number of numerical studies [55, 56, 39, 40, 67, 1] are dedicated to the analysis of the acceleration process of MHD jets: these works treat the disk as a boundary condition rotating at a Keplerian rate and they control the outflow rate of the jets by imposing the value of the mass-loading parameter k (Eq. 1) along each of the field lines anchored into the lower boundary. Various studies of this kind addressed many of the questions listed above: here I summarize some of the major results obtained on the subject.

Stationary Solutions

The aim of the first works dedicated to the study of the magneto-centrifugal launching of jets was primarily to test the feasibility of stationary solutions recurring to time-dependent two-dimensional axisymmetric (2.5D) simulations. An example of such a solution obtained with a time-dependent MHD code is shown in the left panel of Fig. 1: this type of simulation shows clearly the acceleration of a jet flowing parallel to the magnetic field lines crossing all the critical surfaces. This is basically a confirmation of the mechanism suggested by Blandford and Payne: the magnetic energy stored in the toroidal field at the base of the outflow (last term of the Bernoulli invariant, Eq. 5) both accelerates the plasma along the field lines and increases its angular momentum thus providing a centrifugal acceleration. The great advantage

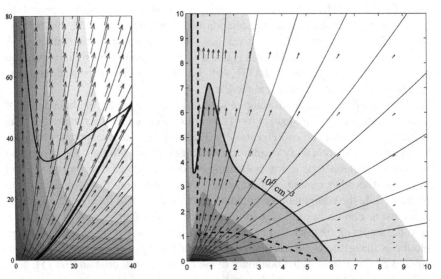

Fig. 1 *Left panel*: Representative simulation of a magneto-centrifugal wind launched from a Keplerian disk [40]. Density isocontours (*shades*), magnetic field lines (*light solid lines*), velocity vectors (*arrows*) and the fast-magnetosonic surface (*solid line of medium thickness*) are plotted. *Right panel*: Similar simulation in which the magnetic flux and the mass-loading are concentrated in an inner region of a protostellar disk [40]. The dashed line marks the fast-magnetosonic surface

of time-dependent simulations with respect to self-similar analytical models is that the Alfvén speed at the disk surface is not obliged to follow the Keplerian scaling $\propto r^{-1/2}$ and therefore the magnetic flux distribution, the density and therefore the k parameter can be chosen independently. An example of a non-self-similar stationary solution obtained with an MHD time-dependent code is shown in the right panel of Fig. 1: here the magnetic and the mass fluxes are concentrated in a tiny region in the central part of the accretion disk. This is somewhat representative of an X-Wind scenario: notice the different degree of collimation and the different shape of the characteristic surfaces of the two cases shown here. It is important to point out that due to its peculiar magnetic flux distribution, no analytical self-similar models are capable of calculating an X-Wind solution. Anyway, it must be kept in mind that in the numerical simulations presented here, the magnetic and dynamical properties of the outflows are not linked with the dynamics of the accretion disk which is not treated self-consistently. Recent studies [24, 59] have examined the relation between the collimation properties of the outflows and the magnetic flux distribution inside the disk.

Non-stationary Solutions and Variability

With time-dependent codes it is possible to study non-stationary solutions, trying to address the fundamental question of variability. Observed T Tauri jets are in fact far from being stationary: their emission knots mark internal working surfaces, i.e., shocks propagating along the jet axis with a typical speed $\sim 50\,\mathrm{km\,s}^{-1}$ with respect to the mean flow [42].

It has been shown for example [1] that the outflow becomes unsteady above some critical mass loss rate (see right panel in Fig. 2): in a "heavy" wind regime the magnetic field is dominated by the toroidal component all the way down to the

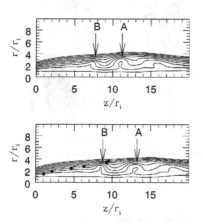
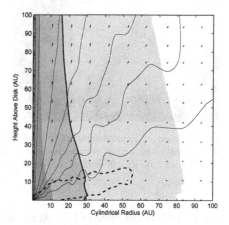

Fig. 2 *Left panel*: Magnetic field structure of an unsteady jet [56]. The capital letters mark the position of the MHD shocks which are forming the knots. *Right panel*: Nonsteady flow behavior of a jet characterized by a high mass outflow rate [1]

launching surface. This magnetic configuration is known to be prone to strong kink instabilities.

The production of knots indicative of MHD shocks propagating along the jet axis had been previously pointed out in [56] (see left panel in Fig. 2): in this case the boundary conditions are probably "overdetermined" (see Sect. 2.2.1), fixing in particular the behavior of the toroidal field at the disk surface. This effect prevents probably the solution to converge towards a stationary state and produces the observed knots. Anyway it is clear that in this type of simulations the non-stationarity and variability of the solutions is a consequence of the imposed boundary conditions at the disk surface: if these conditions can be supported by an actual accretion disk and produce the correct timescales for knot production is unclear.

Three-Dimensional Simulations and Stability

MHD jets dominated by a toroidal field component, which is the case for all the solutions presented, especially beyond the Alfvén surface, are subject to strong non-axisymmetric instabilities. The three-dimensional modeling of the launching process becomes therefore an essential tool to study the stability of this mechanism. Anyway, not many examples of this type of study are available. In particular it has been found in [57] that "corkscrew" or wobbling solutions are not necessarily destroyed by non-axisymmetric ($m = 1$) modes (see the left panel of Fig. 3): this is due to a self-regulatory mechanism which maintains the flow sub-Alfvenic and therefore more stable. In [2] it has been suggested that asymmetric jets can be stabilized by a (light) fast-moving outflow close to the axis with a poloidally dominated magnetic field (see right panel of Fig. 3).

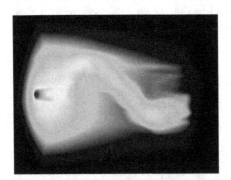 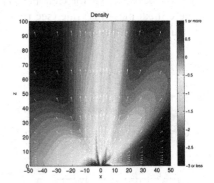

Fig. 3 *Left panel*: Representation of the mass density integrated along the line of sight of a three dimensional simulation of a "corkscrew" jet [57]. *Right panel*: three dimensional simulation of a non-axisymmetric jet stabilized by the presence of a fast-moving central spine [2]

2.2.3 MHD Simulations of the Accretion–Ejection Process

The very first numerical experiments devoted to the study of the launching mechanism of astrophysical jets actually included the disk in the computational domain

[66]. On the other hand, they followed the evolution of the disk–jet system for very short timescales, describing only the transient phase of the evolution of an accretion disk threaded by a large-scale magnetic field ("sweeping magnetic twist").

Inspired by the analytical works by Ferreira and collaborators [25, 26, 12], new numerical studies have been published more recently [14, 15, 75]. As in [26, 12] the diffusive effects inside the accretion disk have been taken into account by parametrizing the disk resistivity with an α prescription (see Sect. 2.1). These works confirmed the results obtained by the analytical modeling. First of all an equipartition magnetic field ($\mu = B^2/4\pi P \sim 1$) is required to obtain robust quasi-stationary solutions. A stronger field would compromise the radial and vertical equilibrium of the disk. On the other hand, a field too low is unable to provide the fast accretion motion, required to allow for a rapid acceleration at the disk surface. As in the stationary self-similar solutions, also a strong turbulent resistivity ($\alpha_m \sim 1$) is required to converge towards a stationary state (see right panel in Fig. 4). The effects of a weaker resistivity are visible in the left panel of Fig. 4: in this case the advection of the magnetic flux towards the central star dominates over the magnetic diffusion. A strong differential rotation along the field lines therefore appears, the footpoint of the line rotating much faster than its opposite. This differential rotation forms a magnetic structure dominated by the toroidal field component propagating vertically along the outflow axis similar to a "magnetic tower". Notice anyway that a robust ejection is obtained even in regions of the parameter space which are not allowed by the stationary solutions.

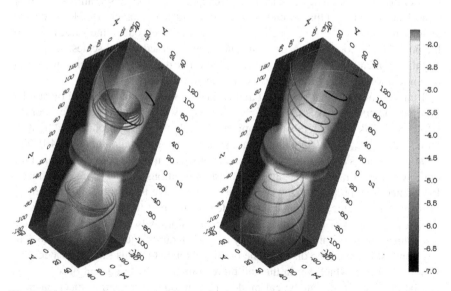

Fig. 4 Three dimensional representation of density for two axisymmetric simulations of a magnetized accretion–ejection structure [75]. In the *left panel* the disk is characterized by a lower resistivity ($\alpha_m = 0.1$) while in the simulation on the right the disk is highly resistive ($\alpha_m = 1$). The *solid lines* represent magnetic *field lines* wrapped around a magnetic surface

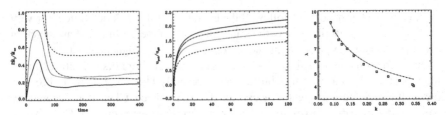

Fig. 5 *Left panel*: Time evolution of the ejection efficiency obtained by simulations characterized by different disk resistivities. Going from the *solid* to the *dashed line* the resistivity parameter decreases from $\alpha_m = 1$ to $\alpha_m = 0.1$. *Central panel*: Vertical profile of the average poloidal speed given in units of the Keplerian speed. Line styles refer to the same cases shown in the *left panel*. *Right panel*: $\lambda - k$ relation calculated along different field lines anchored in the launching region of the simulation corresponding to the *right panel* of Fig. 4. The higher λ values refer to the outer *field lines*. The relation obtained by [73] for a radial wind geometry is plotted with a *dot-dashed line*. All the plots are taken from [75]

The main characteristics of this type of numerical models is summarized in the three panels of Fig. 5. First it is possible to calculate self-consistently the ejection efficiency of the accretion–ejection system: the results obtained ($2\dot{M}_{jet}/\dot{M}_{acc} \sim 0.2$, for the most resistive case) are consistent with observational estimates of this quantity [10]. The terminal speed reaches an asymptotic value around 1–2 times the escape velocity, as shown by observations of T Tauri jets. It is hard to do a direct diagnostic on the rotation speed, since the simulations are able to study a region smaller than the spatial scales currently investigated by HST observations. Anyway it has been shown [28] that extended disk-wind solutions characterized by a specific angular momentum $\lambda \sim 10$ (see Eq. 4), successfully reproduce the values of both the poloidal and toroidal speeds at the currently observed spatial scales. A specific angular momentum $\lambda \sim 9$ is observed in the outer part of the launching region of the simulation shown in the right panel of Fig. 4.

I finally point out that analytical and numerical models characterized by similar disk parameters ($\mu \sim 0.6$, $\alpha_m \sim 1$, $H = 0.1r$) display different jet properties. For the same set of disk parameters a polytropic analytical model [12] produces a "lighter" jet ($k \sim \dot{M}_{jet}/\dot{M}_{acc} \sim 0.01$) with a larger specific angular momentum ($\lambda \sim 40$). The λ parameter gives also a reasonable estimate of the asymptotic poloidal speed of the outflow: $v_{p,\infty} \sim r_0\Omega_K\sqrt{2\lambda - 3}$; the numerical solutions are therefore more mass-loaded and they propagate slower. This is probably due to a problem of numerical diffusion: in the current simulations the density jump at the disk surface, equal to almost four orders of magnitudes in the analytical models [12], is under-resolved. The numerical algorithms tend naturally to smooth out this jump, reducing the density contrast between the disk and the outflow. Notice that this kind of heavier and slower solutions, which are in the end more compatible with current observations, can be obtained in the analytical models by introducing a heating mechanism at the disk surface: this effect increases the mass-loading of the jet by producing a stronger thermal pressure gradient which uplifts the matter from the disk to the outflow [13].

3 Magnetic Star–Disk Interaction

Magnetic fields are likely to play a crucial role also in the dynamics of the very central parts of the star–disk–jet system in T Tauri stars. Polarimetric measures indicate stellar surface magnetic fields up to a few kG [70], with a dipolar or even more complex topology [19, 35]. Such a strong field is able to affect the dynamics of accretion: inverse P-Cygni profiles with strong red-shifted absorption wings are indicative of polar accretion near free-fall velocities along magnetospheric field lines from the inner disk edge [21]. Moreover evidences of the presence of hot stellar winds are starting to emerge, revealed by the broad blue-shifted absorption in high-excitation lines [22, 20]. One of the most puzzling aspects of CTTS is that they are characterized by slow rotation periods (3–10 days, i.e., $< 10\%$ of their break-up speed [7]): this implies a very efficient mechanism of angular momentum removal from the star during its embedded phase. Moreover these stars seem to evolve with a constant rotation speed during the T Tauri phase [34], despite the fact that they are still actively accreting and contracting. The disk truncation and the formation of accretion columns is likely to occur below the corotation radius of the disk, defined as the radius where the stellar angular velocity $\Omega_\star$ is equal to the Keplerian speed of the disk, $R_{co} = \left(GM_\star/\Omega_\star^2\right)^{1/3}$. In this situation, the stellar magnetic field can brake both the disk and the material accreting in the funnel flows. This obviously implies a spin-up torque due to accretion of angular momentum of the order

$$\dot{J}_{\mathrm{acc}} = \dot{M}_{\mathrm{acc}} R_t^2 \Omega_{\mathrm{d}}, \tag{6}$$

where $\dot{M}_{\mathrm{acc}}$ is the accretion rate of the disk, R_t is the truncation radius at which the disk accretion is deviated to form the accretion columns, and Ω_{d} is the disk rotation rate in the truncation region.

Different solutions have been proposed to balance the accretion and the contraction spin-up torques in order to keep the rotation period constant. A first idea is to have an extended magnetosphere connecting the star and the disk beyond the corotation radius: since in this region the disk is rotating slower than the star, angular momentum is transferred from the star back to the disk; if the spin-up and the spin-down torques are equal we find a so-called disk-locked condition [29, 38]. The other solution proposed is based on the presence of outflows, removing the excess angular momentum from the central parts of the system: along this line, it is possible to have stellar winds removing angular momentum directly from the star along the opened field lines of the magnetosphere [49]; another possibility is to have outflows associated with the magnetic star–disk interaction which can remove the disk angular momentum before it falls onto the star. This is the original idea proposed in [65] with the X-Wind scenario.

The problem of the magnetic star–disk interaction and the associated outflows is far too complex to be tackled by analytical means. Analytical techniques can only try to solve single parts of the problem separately, while it is necessary to recur to numerical MHD simulations to have a global view of the system. A brief review of the analytical studies dealing with the star–disk interaction problem will be presented in Sect. 3.1, while the capabilities and results of numerical experiments will be shown in Sect. 3.2.

3.1 Analytical Studies

Analytical studies can aim at solving parts of this complex scenario separately, doing many simplifying assumptions as stationarity and axisymmetry, as in the case of disk-wind models.

One typical and extremely useful example of an analytical study deals with the estimate of the position of the truncation radius R_t. The typical order-of-magnitude estimate related to the accretion onto dipolar magnetospheres is the Alfvénic radius [23]

$$R_A = \left(\frac{B_\star^4 R_\star^{12}}{2GM_\star \dot{M}} \right)^{1/7} ,$$ (7)

derived by equating the ram pressure of a free-falling spherical envelope with the magnetic pressure of the magnetosphere. In this formula $B_\star$ is the intensity of the dipolar field at the stellar surface, $R_\star$ is the stellar radius, $M_\star$ its mass, and $\dot{M}$ is the accretion rate. In the case of a non-spherical accretion, as in the case of an accretion disk, the incertitude on the position of the truncation radius is expressed as $R_t = k R_A$: many authors agree in estimating the proportionality factor around $k \sim$ 0.5−1 [29, 38, 54, 44]. The physics hidden in the factor k has been recently clarified [5]: the proportionality factor can be in fact expressed as $k = (\beta M_s)^{2/7}$, where $\beta = 8\pi P/B^2$ is the ratio of the thermal to magnetic energy and M_s is the sonic Mach number of the accretion flow in the region of the disk close to the truncation radius. The requirement that in the truncation region the dynamics of accretion are dominated by the field leads naturally to the estimates $\beta \sim 1$ and $M_s \sim 1$ and therefore $k \sim 1$.

Other useful studies deal with the structure of magnetically torqued accretion disks [36] and with the angular momentum exchange between the star and a Keplerian accretion disk along a purely dipolar magnetosphere [29, 48]. In particular [48] have shown that the disk-locked scenario is hardly achievable: the angular momentum transfer beyond the corotation radius is probably very inefficient, due to the limited size of the magnetospheric region connecting the star and the disk beyond corotation.

The structure of accretion columns has been studied by many authors based on stationary, axisymmetric, and sub-Alfvénic models of accretion along dipolar magnetic tubes [54, 37]. In particular [37] have shown the crucial role played by the thermal pressure gradient to mass–load the accretion funnels and cross the slow-magnetosonic point. This enhanced pressure gradient can naturally arise from the adiabatic compression of the accretion disk against the "wall" represented by the stellar magnetic field.

Self-similar techniques can be applied to obtain stationary solutions of stellar winds: the meridionally self-similar models calculated by [63] are applicable to study the stellar outflows close to the rotation axis of the star. One major complication in the modeling of stellar winds in the case of slowly rotating young stars, such as T Tauri stars, is that these flows can hardly exploit the protostellar rotational energy and must be therefore essentially driven by their pressure gradient. Such a gradient can come from thermal effects [63] or from turbulent Alfvén waves [18].

3.2 Numerical Simulations

The complexity of the problem of the magnetic star–disk interaction demands the use of heavy numerical experiments. Even in the simplest formulation of the problem, i.e., axisymmetric with a purely dipolar stellar magnetosphere aligned with its rotation axis, global numerical models of the system are not easy to obtain. In particular, many of the first numerical experiments of accretion onto a dipolar magnetosphere did not use strong enough magnetic fields to observe the formation of accretion funnels and relevant magnetospheric outflows: the magnetospheric structure was completely destroyed by the ram pressure of the disk and accretion proceeded directly onto the star surface along the midplane of the disk [33, 53, 41]. The first simulations adopting a strong enough dipolar stellar field to show clearly the formation of accretion funnels emerging at the truncation radius of an accretion disk were published by [60]. After the problem of the formation of accretion columns, the same research group studied the magnetic coupling between the star and the disk [44], while in recent years they were able to perform three-dimensional numerical simulations of accretion onto dipolar [61, 62] and multi-polar [45] magnetospheres inclined with respect to the rotation axis of the star–disk system.

In the following Sections I will review some published and unpublished numerical models of star–disk magnetic interaction. Section 3.2.1 is dedicated to the formation of accretion funnels and the disk-locking paradigm. In Sect. 3.2.2 I will discuss the problem of stellar winds from T Tauri stars. In Sect. 3.2.3 I will show some results showing non-stationary magnetospheric ejections. Finally In Sect. 3.2.4 I will show current state-of-the-art three-dimensional models of magnetic star–disk interaction.

3.2.1 Accretion Funnels and the Disk-Locking Paradigm

The numerical models presented for the first time in [60] showed clearly that a dipolar field component around 1 kG is necessary to truncate an accretion disk with a typical accretion rate around $\dot{M}_{\rm acc} = 10^{-8} M_\odot\,{\rm yr}^{-1}$. This result is consistent with the analytical estimates presented in Sect 3.1. By inserting the expression for the proportionality constant given by [5] we obtain in fact:

$$\frac{R_{\rm t}}{R_\star} = 3\,(\beta M_s)^{2/7} \left(\frac{B_\star}{1{\rm kG}}\right)^{4/7} \left(\frac{\dot{M}_{\rm acc}}{10^{-8} M_\odot {\rm yr}^{-1}}\right)^{-2/7} \left(\frac{M_\star}{0.5 M_\odot}\right)^{-1/7} \left(\frac{R_\star}{2 R_\odot}\right)^{5/7}.$$

$$(8)$$

The numerical simulations presented in [5] also confirmed the validity of the physical conditions assumed in Eq. (8). In the magnetically dominated part of the disk ($\beta \leq 1$) the torque acting on the accretion disk produces in fact an almost sonic accretion ($M_s \leq 1$). Moreover in this situation the magnetic field pressure becomes comparable to the ram pressure of the accreting material ($\beta \sim M_s^{-2} \sim 1$) and the accretion flow is strongly slowed down. This sudden decrease of the Mach number compresses and adiabatically heats the disk: the thermal pressure is now sufficient to uplift the material from the disk and mass-load the accretion columns. Notice that a necessary requirement to obtain a magnetically controlled accretion is that the magnetically dominated part of the disk with $\beta \sim 1$ is located below the corotation radius: if the condition $\beta \sim 1$ is satisfied beyond this radius, the magnetic torque will dominate with respect to the internal viscous torques and spin-up the Keplerian disk rotation: the star–disk system will be therefore in a so-called *propeller* regime, during which only episodic accretion events are possible [68].

Despite a reasonable number of published papers shows nowadays the formation of accretion funnels onto the surface of the star, a numerical simulations showing at the same time the production of the funnels and a magnetosphere extending beyond the corotation radius has never been published. Such a model is crucial to test the validity of the disk-locking paradigm. Both numerical and analytical works have shown that, as the twist due to the differential rotation between the star and the disk increases, more and more toroidal field is generated at the disk surface. When a critical angle $\sim\pi$ is attained, the toroidal field reaches a maximum and its pressure inflates the magnetosphere and opens up the magnetic field structure, thus disconnecting the star and the disk. Anyway it has been shown by [69] that, if the disk resistivity is high enough, thus allowing some azimuthal slippage of the field lines relative to the disk material, this critical angle is never reached and the magnetosphere stays connected with the disk even beyond the corotation.

An unpublished example of a highly resistive and viscous disk interacting with a dipolar stellar magnetosphere is shown in Fig. 6. Both viscosity and resistivity are parametrized recurring to an α prescription [64]: $\nu_{v,m} = \alpha_{v,m} C_s H$, where C_s is the sound speed calculated at the midplane of the disk and H is its thermal heightscale. The connected magnetosphere extends in fact beyond the corotation radius, marked here by the dot-dashed magnetic surface. This in an ideal case to test the disk-locking scenario: the temporal behavior of the torques acting on the surface of the star for this simulation are shown in the right panel of Fig. 7: unfortunately the torque calculated along the magnetosphere closing onto the disk, which include both the accretion and the extended magnetosphere torques, is still spinning up the star rotation. As argued by [48] it is really hard to extend the magnetosphere far enough beyond the corotation radius to balance the torque coming from accretion. The quasi-periodic oscillations showed by the magnetospheric torque reflect the oscillations of the accretion rate (left panel of Fig. 7) with a period around ~ 2 periods of the star: this is a consequence of an unbalance between the viscous torque, controlling the accretion rate on the large scale of the disk and the magnetic torque associated with the star rotation, regulating the accretion in the central region of the disk.

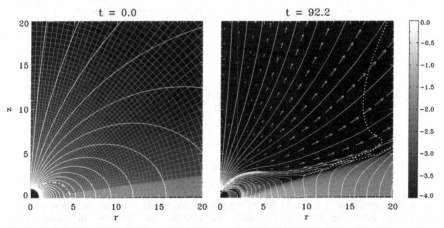

Fig. 6 *Left panel*: Typical initial conditions of an axisymmetric simulation of the interaction of an accretion disk with a dipolar stellar magnetosphere. The disk includes resistive and viscous effects, in order to ensure a continuous accretion inside the disk. The magnetospheric *field lines* are co-rotating with the central star which is modeled as a perfect conductor rotating at 10% of its break-up speed. For a standard normalization ($M_\star = 0.5 M_\odot$, $R_\star = 2 R_\odot$) this corresponds to a period of rotation of 4.5 days. *Right panel*: outcome of a simulation characterized by a stellar field $B_\star = 800$ G and the viscous and resistive α coefficients $\alpha_v = \alpha_m = 1$. The snapshot is taken after $\sim$92 periods of rotation of the protostar. The *dot-dashed line* marks the magnetic surface anchored at the corotation radius, *arrows indicate* the velocity vectors, and the Alfvénic surface of the stellar wind is marked by a *dotted line*. The simulations shown here and in Fig. 9 are performed with the PLUTO code [52]

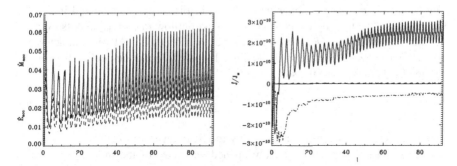

Fig. 7 *Left panel*: Time evolution of the mass accretion rate (*solid line*) and the energy flux (*dashed line*) measured on the surface of the star. For a standard YSO normalization, an accretion rate 0.03 corresponds approximatively to $1.2 \times 10^{-8}\ M_\odot\ \mathrm{yr}^{-1}$. *Right panel*: Time evolution of torques acting on the surface of the star, normalized over the stellar angular momentum. This defines the inverse of the braking time. Plotted are the torques acting along the *opened field lines* (*dot-dashed line*) and on the magnetosphere connected to the disk (*solid line*). Conventionally a positive value spins up the stellar rotation. A value 2×10^{-10} corresponds approximatively to $10^{-6} \mathrm{Myr}^{-1}$. The plots refer to the simulation shown in Fig 6. Time is given in units of the rotation period of the central star.

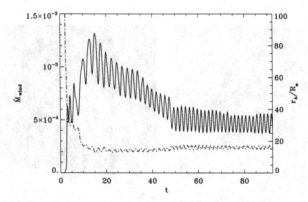

Fig. 8 Time evolution of the mass flux of the stellar wind observed in the simulation shown in Fig. 6. The outflow rate is measured on the *opened filed lines* emerging from the stellar surface (*solid line, left scale*). The average magnetic lever arm of the wind is also plotted (*dot-dashed line, right scale*)

In the right panel of Fig. 7 we can also see that the torque calculated along the opened field lines of the magnetosphere is efficiently spinning down the star rotation: this reflects the presence of a stellar wind flowing along these magnetic surfaces, as clearly visible in Fig. 6. This outflowing component is discussed and described in the following section.

3.2.2 Stellar Winds

The characteristics of the stellar wind shown in Fig. 6 are summarized in Fig. 8, where I show the temporal evolution of its mass outflow rate $\dot{M}_{\mathrm{wind}}$ and its average magnetic lever arm $r_{\mathrm{A}}/R_{\star}$, i.e., the average position of the Alfvén characteristic surface normalized over the stellar radius. These two quantities define the torque exerted by the wind onto the surface of the star:

$$\dot{J}_{\mathrm{wind}} = \dot{M}_{\mathrm{wind}} r_{\mathrm{A}}^2 \Omega_{\star}. \tag{9}$$

The wind torque is strongly enhanced by the huge lever arm $r_{\mathrm{A}}/R_{\star} \sim 16$, which is also a consequence of the widely opened geometry of the magnetic surfaces. This lever arm is basically the same required in [49] to model an "accretion powered" stellar wind capable to balance the spin-up torque coming from accretion and contraction. On the other hand the authors required also an ejection efficiency $\dot{M}_{\mathrm{wind}}/\dot{M}_{\mathrm{acc}} \sim 0.1$, i.e., ten times higher than the one characterizing this simulation, which therefore displays a ten times less efficient torque. A systematic numerical study of "accretion powered" stellar winds has been presented recently by the same authors [50].

As already anticipated in Sect. 3.1, the rotational energy of slowly rotating T Tauri stars is not enough to thrust the stellar wind. An additional energy input at the base of the flow comparable to the gravitational potential energy of the star is needed to give the initial drive to the outflow. In [49] it has been suggested that this extra input of energy can be linked to the energy deposited by the accretion. In the simulation presented here the initial thrust comes from a thermal pressure gradient: on the other hand the temperatures required to launch this type of outflow pose a serious cooling problem. Notice that even if the initial thrust comes from a thermal driving, the outflow is magnetically dominated: the transport of energy and angular momentum is in fact mainly given by the Poynting flux. The magnetic energy could be possibly transferred to the outflowing material beyond the Alfvén surface. The effect of these magnetically dominated stellar winds is similar to the "magnetic towers" propagating along the axis of rotation of the star, introduced by [44] to contribute to the braking torques spinning down the star rotation.

It is therefore clear that, even if stellar winds cannot explain all the dynamical features of T Tauri jets, they can play a crucial role to regulate the rotation period of the star.

3.2.3 Magnetospheric Ejections

Most of the numerical models show that the evolution of a closed magnetosphere connecting a rotating star with a surrounding Keplerian accretion disk leads to the inflation and opening of the magnetospheric field lines. As discussed in Sect. 3.2.1 this is the result of the differential rotation between the star and the disk, and the effect becomes more evident when the magnetic coupling between the disk and the magnetic field is stronger, i.e., when its resistivity is low. The opening of the magnetospheric field lines is accompanied by magnetic reconnection events with plasmoids launched along the current sheet separating the opened field lines anchored inside the disk and the opened field lines anchored onto the star surface (see Fig. 9). Due to the episodic and unsteady character of these ejection events an analytical

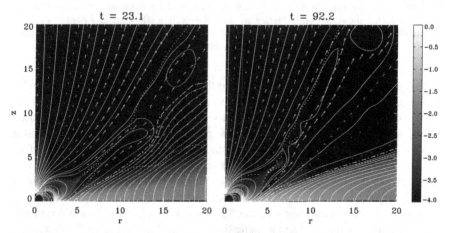

Fig. 9 Time evolution of density maps for a simulation characterized by $\alpha_v = 1$, $\alpha_m = 0.1$, $B_* = 800$ G, $\Omega_* = 0.1\sqrt{GM_*/R_*^3}$. Time is given in units of the stellar rotation period. The ejection of plasmoids due to the inflation and opening of the magnetosphere is visible

analysis of these outflows is not feasible; on the other hand in all the simulation published the magnetic reconnection is controlled by numerical dissipation, thus arising issues on the timescales and the dynamics associated with these phenomena. Moreover, moving ballistically at 45° and not being confined by any external agent, these episodic ejections are not a good candidate to explain the dynamical features of T Tauri jets.

On the other hand the unsteady magnetospheric ejections can represent an efficient mechanism to remove angular momentum from the central parts of the system, thus helping the spin-down of the star. The initial acceleration happens along field lines which are still connecting the star with the disk: the angular momentum is extracted therefore both from the star and the disk, accumulated on the tip of the highly deformed magnetospheric line and then released in the reconnection event, as in a huge magnetic sling.

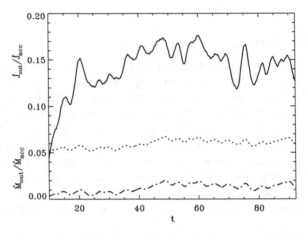

Fig. 10 Time evolution of quantities characterizing the episodic magnetospheric ejections visible in the simulation shown in Fig. 9. Plotted are the mass ejection efficiency (*dot-dashed line*) and the ratio between the ejection torque and the accretion torque (*solid line*). The "Keplerian" torque necessary to accrete from one Keplerian orbit to another is plotted with a *dotted line*

To quantify the angular momentum and the mass extracted from the system by these episodic events is not an easy task. As it is shown in Fig. 10 even when these outflows are characterized by a small mass ejection efficiency the torque exerted on the accretion disk is stronger than a Keplerian torque, i.e., the torque needed to accrete from one Keplerian orbit to another. This type of outflows can therefore be an efficient mechanism to remove angular momentum from the disk *before* it is accreted onto the star. This effect goes back to the original idea of the X-Wind, even if these magnetospheric ejections cannot be characterized as a disk-wind.

This is not the unique magnetic configuration which leads to reconnection and episodic outflows. If a vertical magnetic field is present inside the disk with a polarity which is parallel to the protostellar magnetic moment, a reconnection X-point and a neutral line form at the disk midplane. According to this scenario, accreted

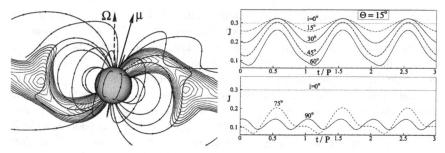

Fig. 11 *Left panel*: Example of three-dimensional simulation describing the interaction between an accretion disk and a magnetosphere with a dipolar momentum inclined with an angle $\theta = 15°$ with respect to the rotation axis of the star. *Right panel*: Synthetic *light curves* obtained from the same simulation corresponding to different viewing angles i

mass is lifted vertically above the X-point by the strong Lorentz force and is loaded onto newly opened field lines anchored onto the stellar surface. If the X-point is located farther than the corotation radius the matter loaded onto the opened field lines is accelerated by the star rotation and produces a so-called "reconnection X-Wind" [27]. No numerical experiments exist of such a magnetic configuration: this would require the simultaneous modeling of the disk magnetic field, and therefore of disk–winds, and of the star–disk magnetospheric interaction.

Numerical models of multi-component outflows including extended disk winds and outflows emerging from the inner part of the system are starting to appear [51], but none of them is able to take into account the dynamics of the star–disk interaction.

3.2.4 Three-Dimensional Simulations

The rotation modulation of the light curves of T Tauri stars, beside giving a measure of the rotation period of the protostar [7], indicates that magnetospheric accretion does not happen along axisymmetric curtains, but probably proceeds along non-axisymmetric columns terminating with accretion spots on the surface of the star. Moreover, the photometric and spectroscopic variations of AA Tau [8] are probably determined by the periodic occultations of a disk warped by the interaction with a dipolar magnetosphere inclined with respect the rotation axis of the star. Sectropolarimetric measurements indicates that the magnetic field of CTTS include multipolar components other than purely dipolar [19, 35]. These observations put in evidence the need for three dimensional simulations to develop models closer to a realistic situation.

To date, only one research group has attacked such a challenging problem [61, 62]. They have been able to model the interaction between a viscous accretion disk and a dipolar magnetosphere with the magnetic moment inclined with respect the rotation axis of the star–disk system (Fig. 11). By varying the inclination angle, they showed how two stream funnel flows develop for mild inclinations ($\theta \sim 15°$)

while it is possible to have direct accretion onto the poles of the star for higher angles ($\theta \sim 60°$). As suggested by the AA Tau observations, the disk is also warped perpendicularly to the magnetic moment in the direction opposed to the position of the accretion columns. The accretion rate shows also a dependency on the inclination angle, being higher for greater θ, when accretion proceeds more directly onto the stellar poles. By estimating the emissivity of the accretion spots with the kinetic energy dissipated by the accretion columns onto the surface of the star, it is also possible to produce synthetic light curves: they showed how it is possible to have only one peak of intensity during one orbital period if the angle of view is smaller than $i < 60°$; the two spots corresponding to the two opposed accretion columns are visible for greater viewing angles. In more recent developments of their work they also took into account multipolar magnetospheres [45] and more recently they started also to perform radiative transfer calculations along the accretion columns.

Unfortunately, due to the small density contrast between the disk and the surrounding atmosphere, they were not able to study the ejection phenomena associated with this complex system: in their three dimensional simulations the star is always spun-up by the star–disk interaction. The problem of the regulation of the stellar period of rotation remains therefore an opened and debated issue.

Nevertheless, it is clear that this kind of simulations represent the more challenging but correct approach to study the star–disk magnetospheric interaction in a situation as much realistic as possible.

4 Summary and Perspectives

The overview presented here has shown that numerical time-dependent simulations represent a powerful tool to study the dynamics of the central regions of accreting T Tauri stars.

Contrary to stationary analytical models, the simulations of the MHD launching mechanism of jets from Keplerian disks are not subject to self-similarity constraints, and they can take into account time-dependent and three-dimensional effects. On the other hand, despite giving results which are in good agreement with observations of T Tauri jets, the numerical experiments considering also the disk dynamics and the global accretion–ejection process are still affected by important numerical effects, such as diffusion and numerical heating. An investment in computing power (resolution) and algorithm development is strongly required.

Numerical MHD simulations are the only tool available to study the interaction between an accretion disk and the stellar magnetosphere. The complex dynamics of the star–disk interaction encompasses accretion and ejection phenomena which are causally linked and need to be studied by global numerical models. In this review I showed in particular that outflows (stellar winds, magnetospheric ejections) play a crucial role in balancing the contraction and the accretion torques which tend to spin-up the stellar rotation.

The main objectives and the future of numerical simulations is to include physical ingredients to produce more realistic models to be compared with observations:

three-dimensional models, including radiative transfer, chemistry, and realistic heating-cooling mechanisms. One warning should be always kept in mind: simulations are a *numerical* and therefore *approximate* solutions of the system of equations that must be solved. Numerical spurious effects are always at work and can sometimes seriously taint the results of the simulations.

References

1. Anderson, J.M., Li, Z.-Y., Krasnopolsky, R., Blandford, R.D., 2005, The structure of magnetocentrifugal winds. I. Steady mass loading. Astrophys J 630, 945–957
2. Anderson, J.M., Li, Z.-Y., Krasnopolsky, R., Blandford, R.D., 2006, Magnetocentrifugal winds in three dimensions: A nonaxisymmetric steady state. Astrophys J 653, L33–L36
3. Bacciotti, F., Eislöffel, J., 1999, Ionization and density along the beams of Herbig-Haro jets. Astron Astophys 342, 717–735
4. Bacciotti, F., Ray, T.P., Mundt, R., Eislöffel, J., Solf, J., 2002, Hubble space telescope/STIS spectroscopy of the optical outflow from DG Tauri: Indications for rotation in the initial jet channel. Astrophys J 576, 222–231
5. Bessolaz, N., Zanni, C., Ferreira, J., Keppens, R., Bouvier, J., 2008, Accretion funnels onto weakly magnetized young stars. Astron Astophys 478, 155–162
6. Blandford, R.D., Payne, D.G., 1982, Hydromagnetic flows from accretion discs and the production of radio jets. MNRAS 199, 883–903
7. Bouvier, J., Wichmann, R., Grankin, K. et al., 1997, COYOTES IV: The rotational periods of low-mass Post-T Tauri stars in Taurus. Astron Astophys 318, 495–505
8. Bouvier, J., Grankin, K.N., Alencar, S. et al., 2003, Eclipses by circumstellar material in the T Tauri star AA Tau. II. Evidence for non-stationary magnetospheric accretion. Astron Astophys 409, 169–192
9. Cabrit, S., Edwards, S., Strom, S.E., Strom, K.M., 1990, Forbidden-line emission and infrared excesses in T Tauri stars – Evidence for accretion-driven mass loss? Astrophys J 354, 687–700
10. Cabrit, S., 2007, The accretion-ejection connection in T Tauri stars: Jet models vs. observations. IAU Symposium 243, 203–214
11. Cai, M.J., Shang, H., Lin, H.-H., Shu, F.H., 2008, X-Winds in Action. Astrophys J 682, 489–503
12. Casse, F., Ferreira, J., 2000, Magnetized accretion-ejection structures. IV. Magnetically-driven jets from resistive, viscous, Keplerian discs. Astron Astophys 353, 1115–1128
13. Casse, F., Ferreira, J., 2000, Magnetized accretion-ejection structures. V. Effects of entropy generation inside the disc. Astron Astophys 361, 1178–1190
14. Casse, F., Keppens, R., 2002, Magnetized accretion-ejection structures: 2.5- dimensional Magnetohydrodynamic simulations of continuous ideal jet launching from resistive accretion disks. Astrophys J 581, 988–1001
15. Casse, F., Keppens, R., 2004, Radiatively inefficient magnetohydrodynamic accretion-ejection structures. Astrophys J 601, 90–103
16. Coffey, D., Bacciotti, F., Woitas, J., Ray, T.P., Eislöffel, J., 2004, Rotation of jets from young stars: New clues from the hubble space telescope imaging spectrograph. Astrophys J 604, 758–765
17. Contopoulos, J., Lovelace, R.V.E., 1994, Magnetically driven jets and winds: Exact solutions. Astrophys J 429, 139–152
18. DeCampli, W.M., 1981, T Tauri winds. Astrophys J 244, 124–146
19. Donati, J.-F., Jardine, M.M., Gregory, S.G. et al., 2008, Magnetospheric accretion on the T Tauri star BP Tauri. MNRAS 386, 1234–1251
20. Dupree, A.K., Brickhouse, N.S., Smith, G.H., Strader, J., 2005, A hot wind from the classical T Tauri Stars: TW Hydrae and T Tauri. Astrophys J 625, L131–L134

21. Edwards, S., Hartigan, P., Ghandour, L., Andrulis, C., 1994, Spectroscopic evidence for magnetospheric accretion in classical T Tauri stars. Astron J 108, 1056–1070

22. Edwards, S., Fischer, W., Kwan, J., Hillenbrand, L., Dupree, A.K., 2003, He I λ 10830 as a Probe of Winds in Accreting Young Stars. Astrophys J 599, L41–L44

23. Elsner, R.F., Lamb, F.K., 1977, Accretion by magnetic neutron stars. I – Magnetospheric structure and stability. Astrophys J 215, 897–913

24. Fendt, C., 2006, Collimation of Astrophysical Jets: The role of the accretion disk magnetic field distribution. Astrophys J 651, 272–287

25. Ferreira, J., Pelletier, G., 1995, Magnetized accretion-ejection structures. III. Stellar and extragalactic jets as weakly dissipative disk outflows. Astron Astophys 295, 807–832

26. Ferreira, J., 1997, Magnetically-driven jets from Keplerian accretion discs. Astron Astophys 319, 340–359

27. Ferreira, J., Pelletier, G., Appl, S., 2000, Reconnection X-winds: Spin-down of low-mass protostars. MNRAS 312, 387–397

28. Ferreira, J., Dougados, C., Cabrit, S., 2006, Which jet launching mechanism(s) in T Tauri stars?. Astron Astophys 453, 785–796

29. Ghosh, P., Lamb, F.K., 1979, Accretion by rotating magnetic neutron stars. III - Accretion torques and period changes in pulsating X-ray sources. Astrophys J 234, 296–316

30. Goodson, A.P., Winglee, R.M., Böhm, K.-H., 1997, Time-dependent accretion by magnetic Young Stellar Objects as a Launching Mechanism for stellar jets. Astrophys J 489, 199–209

31. Gracia, J., Vlahakis, N., Tsinganos, K., 2006, Jet simulations extending radially self-similar magnetohydrodynamics models. MNRAS 367, 201–210

32. Hartigan, P., Edwards, S., Ghandour, L., 1995, Disk accretion and mass loss from young stars. Astrophys J 452, 736–768

33. Hayashi, M.R., Shibata, K., Matsumoto, R., 1996, X-Ray flares and mass outflows driven by magnetic interaction between a protostar and its surrounding disk. Astrophys J 468, L37–L40

34. Irwin, J., Hodgkin, S., Aigrain, S. et al., 2007, The Monitor project: Rotation of lowmass stars in the open cluster NGC2516. MNRAS 377, 741–758

35. Jardine, M.M., Gregory S.G., Donati, J.-F., 2008, Coronal structure of the classical T Tauri star V2129 Oph. MNRAS 386, 688–696

36. Kluźniak, W., Rappaport, S., 2007, Magnetically Torqued thin accretion disks. Astrophys J 671, 1990–2005

37. Koldoba, A.V., Lovelace, R.V.E., Ustyugova, G.V., Romanova, M.M., 2002, Funnel flows from disks to magnetized stars. Astrophys J 123, 2019–2026

38. Königl, A., 1991, Disk accretion onto magnetic T Tauri stars. Astrophys J 370, L39–L43

39. Krasnopolsky, R., Li, Z.-Y., Blandford, R., 1999, Magnetocentrifugal launching of jets from accretion disks. I. Cold Axisymmetric Flows. Astrophys J 526, 631–642

40. Krasnopolsky, R., Li, Z.-Y., Blandford, R., 2003, Magnetocentrifugal Launching of Jets from Accretion Disks. II. Inner Disk-driven Winds. Astrophys J 595, 631–642

41. Küker, M., Henning, T., Rüdiger, G., 2003, Magnetic star-disk coupling in classical T Tauri systems. Astrophys J 589, 397–409

42. Lavalley–Fouquet, C., Cabrit, S., Dougados, C., 2000, DG Tau: A shocking jet. Astron Astophys 356, L41–L44

43. Li, Z.-Y., 1996, Magnetohydrodynamic disk-wind connection: Magnetocentrifugal winds from ambipolar diffusion-dominated accretion disks. Astrophys J 465, 855–868

44. Long, M., Romanova, M.M., Lovelace, R.V.E., 2005, Locking of the rotation of disk-accreting magnetized stars. Astrophys J 634, 1214–1222

45. Long, M., Romanova, M.M., Lovelace, R.V.E., 2008, Three-dimensional simulations of accretion to stars with complex magnetic fields. MNRAS 386, 1274–1284

46. Matsakos, T., Tsinganos, K., Vlahakis, N. et al., 2008, Two-component jet simulations. I. Topological stability of analytical MHD outflow solutions. Astron Astophys 477, 521–533

47. Matt, S., Goodson, A.P., Winglee, R.M., Böhm, K.-H., 2002, Simulation-based Investigation of a Model for the interaction between stellar magnetospheres and circumstellar accretion disks. Astrophys J 574, 232–245

48. Matt, S., Pudritz, R.E., 2005, The spin of accreting stars: dependence on magnetic coupling to the disc. MNRAS 356, 167–182
49. Matt, S., Pudritz, R.E., 2005, Accretion-powered stellar winds as a solution to the stellar angular momentum problem. Astrophys J 632, L135–L138
50. Matt, S., Pudritz, R.E., 2008, Accretion-powered stellar winds. II. Numerical solutions for stellar wind torques. Astrophys J 678, 1109–1118
51. Meliani, Z., Casse, F., Sauty, C., 2006, Two-component magnetohydrodynamical outflows around young stellar objects. Interplay between stellar magnetospheric winds and disc-driven jets. Astron Astophys 460, 1–14
52. Mignone, A., Bodo, G., Massaglia, S. et al., 2007, PLUTO: A numerical code for computational astrophysics. Astrophys J Suppl Ser 170, 228–242
53. Miller, K.A., Stone, J.M., 1997, Magnetohydrodynamic simulations of stellar magnetosphere–accretion disk interaction. Astrophys J 29, 890–902
54. Ostriker, E.C., Shu, F.H., 1995, Magnetocentrifugally driven flows from young stars and disks. IV. The accretion funnel and dead zone. Astrophys J 447, 813–828
55. Ouyed, R., Pudritz, R.E., 1997, Numerical simulations of astrophysical jets from keplerian disks. I. stationary models. Astrophys J 482, 712–732
56. Ouyed, R., Pudritz, R.E., 1997, Numerical simulations of astrophysical jets from keplerian disks. II. Episodic outflows. Astrophys J 484, 794–809
57. Ouyed, R., Clarke, D.A., Pudritz, R.E., 2003, Three-dimensional Simulations of Jets from Keplerian Disks: Self-regulatory Stability. Astrophys J 582, 292–319
58. Pelletier, G.: *Introduction to Magneto-Hydrodynamics*. Lect. Notes Phys. **723**, 77–101. Springer Verlag, Berlin (2007)
59. Pudritz, R.E., Rogers, C.S., Ouyed, R., 2006, Controlling the collimation and rotation of hydromagnetic disc winds. MNRAS 365, 1131–1148
60. Romanova, M.M., Ustyugova, G.V., Koldoba, A.V., Lovelace, R.V.E., 2002, Magnetohydrodynamic simulations of disk-magnetized star interactions in the quiescent regime: Funnel flows and angular momentum transport. Astrophys J 578, 420–438
61. Romanova, M.M., Ustyugova, G.V., Koldoba, A.V., Wick, J.V., Lovelace, R.V.E., 2003, Three-dimensional simulations of disk accretion to an inclined dipole. I. Magnetospheric flows at different Θ. Astrophys J 595, 1009–1031
62. Romanova, M.M., Ustyugova, G.V., Koldoba, A.V., Lovelace, R.V.E., 2004, Three-dimensional simulations of disk accretion to an inclined dipole. II. Hot spots and variability. Astrophys J 610, 920–932
63. Sauty, C., Trussoni, E.,Tsinganos, K., 2002, Nonradial and nonpolytropic astrophysical outflows. V. Acceleration and collimation of self-similar winds. Astron Astophys 389, 1068–1085
64. Shakura, N.I., Sunyaev, R.A., 1973, Black holes in binary systems. Observational appearance. Astron Astophys 24, 337–355
65. Shu, F., Najita, J., Ostriker, E., Wilkin, F., Ruden, S., Lizano, S., 1994, Magnetocentrifugally driven flows from young stars and disks. 1: A generalized model. Astrophys J 429, 781–796
66. Uchida, Y., Shibata, K., 1985, Magnetodynamical acceleration of CO and optical bipolar flows from the region of star formation. PASJ 37, 515–535
67. Ustyugova, G.V., Koldoba, A.V., Romanova, M.M., Chechetkin, V.M., Lovelace, R.V.E., 1999, Magnetocentrifugally driven winds: Comparison of MHD simulations with theory. Astrophys J 516, 221–235
68. Ustyugova, G.V., Koldoba, A.V., Romanova, M.M., Lovelace, R.V.E., 2006, "Propeller" Regime of disk accretion to rapidly rotating stars. Astrophys J 646, 304–318
69. Uzdensky, D.A., Königl, A., Litwin, C., 2002, Magnetically linked star-disk systems. I. Force-free magnetospheres and effects of disk resistivity. Astrophys J 565, 1191–1204
70. Valenti, J.A., Johns-Krull C.M., 2004, Observations of magnetic fields on T Tauri stars. Astrophys Space Sci 292, 619–629
71. Vlahakis, N., Tsinganos, K., 1998, Systematic construction of exact magnetohydrodynamic models for astrophysical winds and jets. MNRAS 298, 777–789

72. Wardle, M., Königl, A., 1993, The structure of protostellar accretion disks and the origin of bipolar flows. Astrophys J 410, 218–238
73. Weber, E.J., Davis L.J., 1967, The angular momentum of the solar wind. Astrophys J 148, 217–227
74. Woitas, J., Bacciotti, F., Ray, T.P., Marconi, A., Coffey, D., Eislöffel, J., 2005, Jet rotation: Launching region, angular momentum balance and magnetic properties in the bipolar outflow from RW Aur. Astron Astophys 432, 149–160
75. Zanni, C., Ferrari, A., Rosner, R., Bodo, G., Massaglia, S., 2007, MHD simulations of jet acceleration from Keplerian accretion disks. The effects of disk resistivity. Astron Astophys 469, 811–828

Jet Stability: A Computational Survey

Rony Keppens, Zakaria Meliani, Hubert Baty, and Bart van der Holst

Abstract To investigate stability properties of astrophysical jets, high-resolution numerical simulations are nowadays used routinely. In this chapter, we address jet stability issues using two complementary approaches: one where highly idealized, classical magnetohydrodynamic (MHD) "jet" configurations are simulated in detail, and one where the full complexity of relativistic jet flows is mimicked computationally. In the former, we collect vital insights into multi-dimensional MHD evolutions, where we start from simple planar, magnetized shear flows to eventually model full three dimensional, helically magnetized jet segments. Such a gradual approach allows an in-depth study of [1] the nonlinear interaction of multiple, linearly unstable modes; as well as [2] their potential to steepen into shocks with intricate shock–shock interactions. All these return to varying degree in the latter approach, where jets are impulsively injected into the simulation domain, and followed over many dynamical timescales. In particular, we review selected recent insights gained from relativistic AGN jet modeling. There, we cover both relativistic hydro and magneto-hydrodynamic simulations. In all these studies, the use of grid-adaptive codes suited for modern supercomputing facilities is illustrated.

R. Keppens (✉)
Centre for Plasma Astrophysics, Leuven Mathematical Modeling and Computational Science Center, K.U.Leuven, Belgium; FOM-Institute for Plasma Physics Rijnhuizen, Nieuwegein, The Netherlands; Astronomical Institute, Utrecht University, The Netherlands, rony.keppens@wis.kuleuver.be

Z. Meliani
Centre for Plasma Astrophysics, K.U.Leuven, Belgium, Zakaria.Meliani@wis.kuleuven.be

H. Baty
Observatoire Astronomique, 11 Rue de l'Université, 67000 Strasbourg, France, baty@astro.u-strasbg.fr

B. van der Holst
University of Michigan, 2455 Hayward Ann Arbor, Michigan 48109-2143, USA, bartvand@umich.edu

Keppens, R. et al.: *Jet Stability: A Computational Survey*. Lect. Notes Phys. **791**, 179–199 (2009)
DOI 10.1007/978-3-642-03370-4_7   © Springer-Verlag Berlin Heidelberg 2009

1 Introduction: Planar MHD Shear Flows

This "computational survey" of jet stability starts with reviewing various results obtained from highly idealized jet models. Using a continuum magnetohydrodynamic viewpoint on jet dynamics, the governing equations express mass, momentum, energy, and magnetic flux conservation in the scale-invariant laws of ideal MHD (an excellent introduction to the Principles of Magnetohydrodynamics is provided in [9]). These form a set of eight nonlinear partial differential equations for the spatio-temporal evolution of density ρ, velocity $\mathbf{v}$, pressure p, and the magnetic field vector $\mathbf{B}$. The latter is subject to Maxwell's equation $\nabla \cdot \mathbf{B} = 0$, expressing the absence of magnetic monopole sources.

1.1 Steady Shock-Dominated MHD Flows

The MHD equations allow for a total of seven wave speeds: the plasma velocity $\mathbf{v}$ advects entropy perturbations, while both forward and backward (with respect to the flow speed) slow, Alfvén, and fast wave modes propagate anisotropically through the magnetized, compressible medium. These linear wave modes are decoupled in a uniform plasma, but may give rise to intricate linear couplings and wave transformations in more realistic configurations. In nonlinear MHD evolutions, the variety of MHD wave modes returns in shock-dominated scenarios, where the various nonlinear counterparts can be obtained in both steady and unsteady evolutions. Especially for configurations involving sizable (shear) flows, complicated shock patterns may arise.

An example for super-fast flow about a cylindrical obstacle is shown in Fig. 1. The planar, up-down symmetric flow modeled there reproduces the stationary, shock-dominated bow shock analyzed in detail by [6, 7]. The defining dimensionless parameters characterizing the flow and magnetic field strength are the Mach number $M = v_x/c_s$ and the plasma beta parameter $\beta = 2p/B^2$, where $c_s = \sqrt{\gamma p/\rho}$ is the sound speed. The bow shock "dimples" as shown in Fig. 1 for inflow where $M = 2.6$ and $\beta = 0.4$. When the inflow varies from flow-dominated to magnetically dominated, a mere parabolic bow shock in front of the obstacle is no longer admissible from symmetry considerations when the inflow is in the switch-on regime. In this regime, a tangential magnetic field component develops downstream, while there is none upstream, hence the name "switch-on". Along the symmetry line, a mere hydrodynamic shock then forms, which connects further on to shock segments of intermediate and fast magnetosonic type.

1.2 Kelvin–Helmholtz Unstable, Magnetized Shear Flows

Perhaps the most well-known effect related to the presence of shear flow is the possibility of Kelvin–Helmholtz instability. A planar shear flow where $v_x \propto \tanh(y/a)$

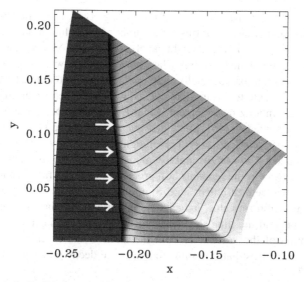

Fig. 1 A magnetically dominated super-fast flow (*arrows*) upstream of a perfectly conducting cylindrical obstacle (*bottom right*) results in a steady bow shock consisting of multiple interacting shock segments. We show the density in greyscale and the magnetic *field lines*, for a case with inflow parameters $M = 2.6$ and $\beta = 0.4$

can be unstable for a fairly wide range of prevailing Mach number and plasma beta. The presence of a uniform, flow-aligned magnetic field reduces the growth rate obtained for a similar pure hydrodynamic flow pattern and one can make a quantitative analysis of this stabilizing behavior by solving the compressible MHD equations, linearized about the stationary shear flow equilibrium. This linearization quantifies the normal modes with temporal behavior $\exp(-i\omega t)$, where the frequency $\omega = \omega_R + i\omega_I$ is complex in general and can adopt a Fourier representation $\exp(i2\pi x/\lambda_x)$ in the streamwise x-direction. Such stability analysis (see, e.g., [26]) predicts that when $M = 1$ and $\beta = 120$, the maximal growth rate ω_I is found when the horizontal wavelength $\lambda_x \approx 20a$, or about 20 times the typical length scale of the cross-stream shear profile. The up-down (a)symmetry makes the phase velocity of this mode vanish, i.e., $\omega_R = 0$.

Extensive literature exists on numerical studies of the nonlinear behavior of linearly unstable, magnetized shear flows [20, 11, 28, 12]. For the mentioned parameter values, the magnetic field gets amplified in the vortical roll-up of the velocity shear layer, in narrow strands about the vortex perimeter. These intersect the compression zones and coincide with localized regions of density depletion. In Fig. 2 (left panel), the density variation is shown near the time of the nonlinear mode saturation, at which time the localized lanes of amplified magnetic field start to control the further vortical deformation. The right panel from Fig. 2 shows how a similar shear-flow configuration, initially coinciding with a co-spatial current sheet where the magnetic field also reverses direction, has an additional route to both linear and nonlinearly induced instability. Accounting for a finite amount of (either numerical or properly resolved) resistivity allows for a sausage type, linear pinching mode as well, where a magnetic island forms. In the nonlinear evolution, that island can be seen at the periodic sides in Fig. 2 (right panel), while additional, smaller islands can form and interact along the low density lane above and below the central "cat's eye" pattern. Since now antiparallel field regions are forced into contact, these current layers become unstable to tearing-like events, which are here induced by the non-linear Kelvin–Helmholtz development. A fairly extensive parameter study of this behavior was presented in [15]. Since the reversed magnetic field case is strongly influenced by resistive effects, a grid resolution study in fully resistive MHD computations was needed to convincingly demonstrate how the magnetic energy can be tapped effectively in the reconnection processes, quickly transforming the shear layer to a more magneto-turbulent state. Significant density deviations of up to 40 % of their original, uniform value, indicate the importance of plasma compressibility. Furthermore, beyond the time shown in Fig. 2, reconnection events are important in the disruption of the vortex, for both the aligned and reversed magnetic field case. More recent investigations allow to assess the decay rate as a function of the resistivity [27].

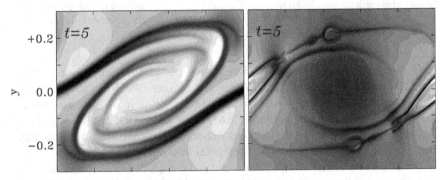

Fig. 2 (Color online) The density pattern obtained at the nonlinear saturation time for a Kelvin–Helmholtz unstable, magnetized shear flow. The *left panel* started from a uniform, flow-aligned magnetic field, while the *right panel* is for an initial reversed field configuration. Times are in units of horizontal sound crossings, and this case has a $\beta \simeq 120$ and is at unit Mach number

2 From Planar Shear Flows to Three Dimensional Jets

Both examples discussed above already demonstrate some of the complexity one can encounter in magnetized shear flow configurations. In what follows, we will gradually relax some of the idealizations that were assumed. In so doing, we revisit results which can help unravel various details that are also present in the more realistic, three dimensional, scale-encompassing astrophysical jet simulations which are typically pursued. In the second part of this survey, some exemplary studies of relativistic jets will be presented as well, where the effects discussed here will return.

2.1 Vortex Disruptions and Mode Interactions

The planar $v_x \propto \tanh y$ velocity profile mentioned thus far focused on the nonlinear evolution of a single Kelvin–Helmholtz vortex or "billow". An in-depth study of the nonlinear mode saturation started by choosing the simulation box size in accord with the wavelength λ_x of the most unstable mode. This obviously excludes the development of subharmonics characterized by double, quadruple, etc. horizontal wavelengths. While the single billow run identified the role of magnetic field reconnections in the vortex disruption beyond the saturation time, it is relevant to extend the domain size for a single shear layer to span multiple streamwise wavelengths of the most unstable mode. This can be done most affordably using grid-adaptive computations, employing Adaptive Mesh Refinement (AMR). With the AMRVAC code [16, 30], both extended shear layers [3] as well as double shear layers [4], have been simulated to quantify the role of mode interactions.

A clear motivation to investigate extended magnetized shear layers comes from pure hydrodynamic studies: it is known that in compressible planar shear flow, a trend to vortex interaction and coalescence is observed. The Kelvin–Helmholtz instability leads to alternating compression and depression zones, which first form at length scales characteristic for the most unstable linear mode. The nonlinear evolution of a transonic $M = 1$ shear layer, with a nearly negligible uniform magnetic field where $\beta = 12,000$, is shown in Fig. 3 in its density evolution. A clear trend to large-scale, ultimately halted by the finite size of the simulation box, is evident. Vortices pair and merge, and the fate of this large-scale trend under more dynamically important magnetic fields is of interest. It was then found that for parameters similar to single billow study from Sect. 1, both a large-scale trend to coalescing vortices and a small-scale trend to individual vortex disruption emerge. The coalescence is a direct consequence of mode–mode interactions, where subharmonics of double wavelength play a prominent role. This can be shown in deterministic runs where the box size is taken exactly twice the wavelength of the most unstable mode. If one perturbs the shear layer with this mode and its subharmonic, pairing results under zero phase difference between both modes. This corresponds to an initial evolution where the two vortices are of equal strength, but get displaced asymmetrically about the original shear layer. Under $\pi/2$ phase difference, both vortices evolve independently, in which case the two vortices are initially of unequal strength while remaining centered on the shear layer.

It is even possible to find an intermediate parameter regime between the hydrodynamic trend and the disruptive MHD scenario. Figure 4 demonstrates consecutive density patterns found for a layer where $\beta \approx 1000$. The first panel shows originally 8 compression–depression zones, which merge in a pure hydrodynamic fashion. After about 11 time units (where the normalization is with respect to a sound crossing time), the initially weak field has survived consecutive roll-up events and the distinctive low density lanes coincide with amplified field regions. Since the field orientation alternates after multiple roll-ups, antiparallel field regions exist and the distinctive tearing events mentioned earlier return at the vortex periphery. Still, the trend to large scale persists, this time complicated by the sudden transition

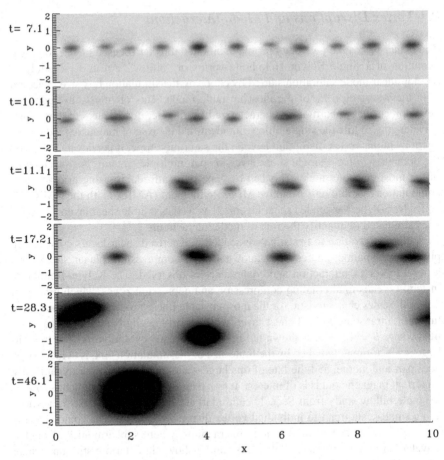

Fig. 3 Density evolution for an extended compressible shear layer, with a very weak magnetic field ($\beta = 12000$). This transonic $M = 1$ layer simulation uses units related to sound crossings

to magneto-turbulent flow. Hence, even when starting from uniform magnetic field configurations, the formation and dynamical role of localized current layers imposes stringent demands on resolution, most adequately achieved using AMR techniques.

2.2 Double Shear Layers

While single shear layers represent flow conditions in a cross-section localized at the boundary of a true three dimensional jet flow, a more jet-relevant flow pattern accounts for the presence of two shear flow layers bounding the jet (again in a planar representation). This brings in one extra dimensionless parameter, measuring the jet width $2R_{\text{jet}}$ in units of the shear layer width a.

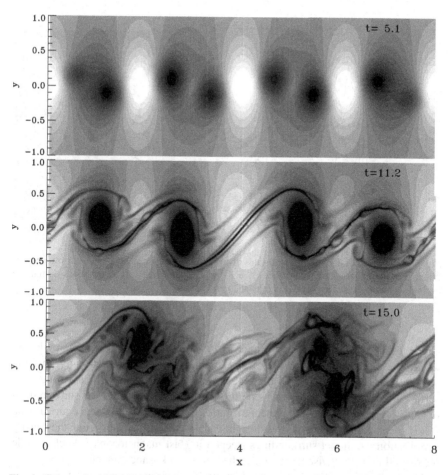

Fig. 4 (Color online) The density evolution for an extended compressible shear layer, at an initial magnetic field strength such that one first obtains hydrodynamic vortex mergers, and eventually the system transits to magneto-turbulent flow. The plasma $\beta = 1000$ and the layer is transonic $M = 1$

One again can resort to linear MHD solvers to quantify the most unstable modes, their growth rates, phase speeds, and eigenfunctions. In the limit of widely separate shear layers, one finds the same result known for individual shear layers, although a degeneracy exists between sinuous (asymmetric in their density eigenfunction with respect to the jet midplane) and varicose (symmetric or sausage like) modes. Under transonic $M = 1$, magnetized conditions similar to the single billow case from Fig. 2, both modes have identical growth rates for $R_{jet} = 10a$ or larger, while the sinuous, kink-like deformation of the jet is the most unstable linear mode at closer layer–layer separation (its growth rate is about twice larger than the varicose mode when $R_{jet} = 2.5a$). A quantification of the growth rates found for a supersonic, narrow planar jet is shown in Fig. 5 as found in [4]. Various kinds of modes are quantified there by their growth rate as function of wave number (both of surface

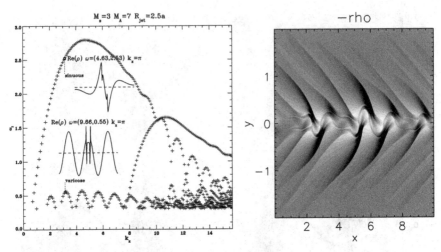

Fig. 5 (Color online) *Left*: The growth rates of unstable modes for a supersonic, narrow jet, as a function of the horizontal wave number $k_x = 2\pi/\lambda_x$. *Right*: the density in the shock-dominated evolution of a planar supersonic Mach 6 jet at about $t = 3.6$

and body type). The density eigenfunction for a selected sinuous surface mode and a varicose mode is shown as well. Supersonic, narrow jets, are strongly unstable to kink-type deformations, and the eigenfunctions already show their strong non-surface like character: compressive perturbations are found in the far surroundings of the jet. A snapshot of the nonlinear evolution of a Mach $M = 6$, near equipartition $\beta = 1.63$ planar jet is shown as well (right panel), and one can detect that the density perturbations in the jet surroundings steepen in (fast magnetosonic) shocks, while the overall jet is kinkwise perturbed with strong shock–shock interactions.

Due to the trend to large scale found on single shear layers mentioned in Sect. 2.1, eventually all planar jet studies will witness layer–layer interactions, even for initially widely separated double layers where the linear modes are fully degenerate. A noteworthy effect of layer–layer studies, again generalizing a known pure hydrodynamic result to weak magnetic field regimes, is shown in Fig. 6. This density snapshot of a study where the plasma beta is initially uniform and set to $\beta = 3000$ demonstrates how vortices that form on separate shear layers can demonstrate so-called Batchelor couplings: counter-rotating planar vortices pair up without merging and leave the jet core. This process is known in planar hydrodynamics and is here slightly modified by the weak magnetic field. The localized strands of density depletions again signal the sites of magnetic field amplification. These planar, magnetized jet studies thus already show a surprising richness in their nonlinear dynamics, where vortices can coalesce, form couples, and magnetic fields induce disruptions, or play a prominent role in the shock-dominated transitions.

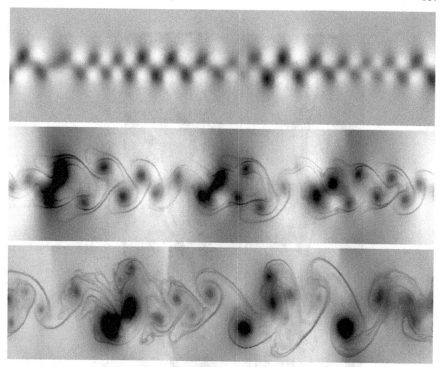

Fig. 6 The density variation showing Batchelor coupling events between counter-rotating vortices forming at opposing ends of a weakly magnetized ($\beta = 3000$) planar jet flow. Times are at approximately 4.5, 6.5, and 9.5 time units (related to sound crossings)

2.3 Wake-Current Sheets

A relevant, closely related study is found in local box computations of the solar coronal streamer belt [31]. As sketched in Fig. 7, the flow and magnetic field configuration above a coronal helmet streamer has its essential variation in the vertical, cross-stream direction: the fast solar wind is bounding the slow wind wake region. Particular to this configuration is the co-spatial location of a current-sheet due to the variation of $B_x(y)$, and a force-free field representation is applicable at larger heliospheric distances, making $\mathbf{B}(y) = (B_x(y), 0, B_z(y))$. The wake flow-current sheet system shares many similarities with the idealized models discussed thus far, and defining parameters set flow and field strength, while an extra parameter can be fixed when taking the current sheet width roughly equal to the shear flow width. A wake flow profile is obtained when $v_x(y) \propto (1 - \cosh^{-1} y)$, and a linear stability study needs to explore the extended parameter space spanned by sonic Mach number, plasma beta (or Alfvénic Mach number), this time allowing for both streamwise $\exp(ik_x x)$ and spanwise $\exp(ik_z z)$ variations. An intriguing finding of such exploration is that despite the essentially one dimensional

Fig. 7 (Color online) A schematic illustration of the local box representing solar coronal streamer belt conditions and the density evolution in a typical case. Times shown are at $t \simeq 110$ and $t \simeq 150$, for a supersonic $M = 3$ wake flow at about unit plasma beta

equilibrium configuration, regions in parameter space exist where the most dominant linear instability has a fully three dimensional character, i.e., both $k_x \neq 0$ as well as $k_z \neq 0$. This is in contrast with incompressible evolutions, where the largest growth rates are always associated with two dimensional modes with $k_z = 0$. The compressible case thus calls for a need to perform three dimensional nonlinear simulations to address the dynamics of wake-current sheets. The right panel of Fig. 7 shows how a super-Alfvénic case representative of the coronal streamer belt conditions again is mostly liable to sinuous deformations, which once more demonstrate density wave patterns reaching into the far surroundings (the fast solar wind region). These similarly steepen into fast magnetosonic shock fronts, pointing to the distinct possibility of in-situ shock formation in the solar corona. These shocks are found for $\beta \simeq 1$ conditions in supersonic $M > 2.6$ wake flows.

2.4 Cylindrical Jets

Returning to jet-like flows, truly three dimensional effects can be analyzed in detail in a case where the flow direction is treated periodically: a segment of a jet flow is then simulated with a resolution otherwise employed in full scale-encompassing simulations. In [14], a cylindrical jet segment with parameters closely resembling the planar single billow study from Fig. 2 was taken. The length of the jet corresponded to a single wavelength for the most unstable mode (so the coalescence effects mentioned thus far are yet to be studied), and transonic $\beta = 120$ conditions start with a uniform flow-aligned magnetic field which pervades the jet and the outer medium.

For specific perturbations of the three dimensional jet, given by their axial (n) and azimuthal (m) integer wave number, predictions could be made regarding the quasilinear mode evolution and excitation due to the initially (relatively) weak magnetization. An $(m, n) = (1, 1)$ perturbation could be shown to induce $(0, 2)$, $(2, 2)$, and $(2, 0)$ modes, in that order. Using both a pseudospectral as well as a finite volume code, these analytic predictions were confirmed in the simulated evolution. Nonlinear effects cause further mode couplings where $m = 3$ deformations are pronounced. These can be understood from an in-depth analysis of the three dimensional variations. Schematically, one finds that

- In a cross-section through the jet axis, the most dominant Kelvin-Helmholtz mode forms at each boundary, doubling the pattern observed in Fig. 2 in an out of phase fashion (hence, of the kink type mentioned in planar studies). Note in Fig. 8 at left the (lightly colored) compression zones, once more containing lanes of low density (dark colored), which extend to low density sheets in three dimensional that curve around the jet circumference.
- The regions of density compression fold around the jet circumference and coincide with the locations of enhanced thermal pressure. The latter correspond to

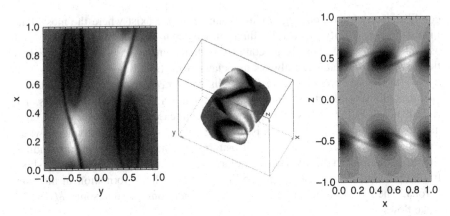

Fig. 8 (Color online) Density pattern in a *horizontal* (*left*) and *vertical* (*right*) cross-section, while a three dimensional rendering of the jet circumference with its thermal pressure variation is shown in the *middle*, all at time $t = 4$ for a $\beta = 120$ transonic jet

the lighter shade regions in the jet isosurface shown in Fig. 8, where it is colored by thermal pressure.

- A cross-cut through the jet axis, orthogonal to the one mentioned above, shows how this induces a wave number doubling in the typical Kelvin–Helmholtz density evolution along the top and bottom of the jet (right panel of Fig. 8).
- The depression zones seen in the latter cross-section extend in a kind of helical manner in three dimensions and together with the sheet-like regions act to locate the regions where magnetic field is amplified locally.
- The localized regions of high magnetic field further control the morphological appearance of the jet deformation: due to both sheetlike and helical fibril regions, a $m = 3$ pattern finally emerges in the jet circumference.

Similar findings can be obtained and understood in detail for $m = 2$ initial perturbations. Ultimately, a breakup of the jet involving many more small-scale wave numbers results in a fairly limited timespan of only six typical sound crossing times. That breakup signals the sudden onset of a more turbulent regime, and the ideal MHD runs typically performed become then influenced by their inherent numerical resistivity.

2.5 Current-Carrying Jets

A final example of a highly idealized, three dimensional MHD simulation recollects findings from [2], where now the local three dimensional jet segment studied above is enriched with a helical initial magnetic field topology. Once more, linear MHD computations allow to identify parameter regimes [1] where not only the surface-like Kelvin–Helmholtz modes are expected to develop, but also additional current-driven modes exist. In fact, varying the field from uniform to twisted under otherwise identical dimensionless parameters (supersonic and axial plasma beta $\beta = 32$), the linear results could be used to contrast the evolution discussed in Sect. 2.4 with one where both modes are unstable. Depending on the twist in the field, the current-driven mode could even become dominant, i.e., having the largest growth rate. The current-driven kink growth rate is indeed naturally controlled by varying the twist of the field, as this ideal instability represents the imbalance of a stabilizing axial tension force from the axial field component, with the destabilizing influence of an azimuthal field component. The latter causes a runaway kink perturbation as its own magnetic pressure variation enhances its growth, as schematically illustrated in the top panel from Fig. 9.

The two linear modes (Kelvin–Helmholtz and current-driven) are easily distinguished in the initial density evolution, as their eigenfunctions show a clearly different character: the $m = -1$ current-driven mode is centrally localized, while the $m = \pm 1$ Kelvin–Helmholtz modes preferentially develop as surface perturbations. A cross-cut through the three dimensional density structure for three cases is shown in Fig. 9 (bottom panel). From left to right, the initial field topology was uniform (as in the previous Sect. 2.4) and increasingly twisted for the rightmost panels. The central deformation of the latter two cases can be traced to the current-driven mode development. Clearly, higher twist cases develop less fine structure in their disruption. This case pointed to the distinct possibility that mode–mode interaction, this time between Kelvin–Helmholtz and current-driven type modes, can aid in jet coherency. The result can be physically motivated as well: the current-driven

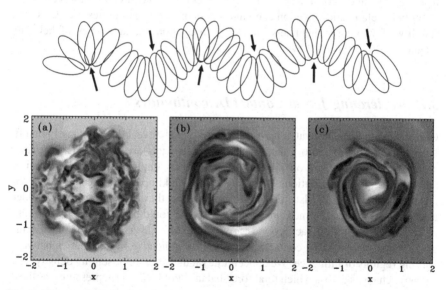

Fig. 9 (Color online) *Top*: a schematic impression of the kink instability arising in a purely azimuthal field topology. *Bottom*: Density pattern in cross-sections through three dimensional jets, starting from uniform (*left*) versus increasingly helical magnetic field scenarios. All are at $t = 14$

development eventually enhances the azimuthal field at the jet periphery, and when this happens during the nonlinear evolution of the Kelvin–Helmholtz vortices, the enhanced tension prevents the Kelvin–Helmholtz vortices to fully disrupt at the jet perimeter.

3 Relativistic AGN Jet Modeling

In the second part of this computational survey on jet stability, we will consider various results obtained with more realistic astrophysical jet models. In particular, astrophysical jets associated with Active Galactic Nuclei (AGN) demonstrate large-scale collimated flows, characterized by relativistic bulk speeds with Lorentz factor of order 10: a large sample of jets showed Lorentz factors between 5 and 30 as inferred from VLBA observations [13]. They are ultimately connected to accretion processes at the center of the radio galaxies, which harbor massive black holes. The jet lengths range from parsec to few hundreds of kiloparsec distances and their radio images form direct evidence for the presence of dynamically important magnetic fields. Indeed, the radio emission is consistent with synchrotron emission from populations of highly relativistic particles. In what follows, we will review three recent modeling results which address stability issues for such relativistic jet flows. All of these adopt a continuum magnetofluid approximation in a flat Minkowski space-time. Hence, they do not address the general relativistic effects at play in the launching regions of these jets, but assume internal jet properties that are plausible from theoretical descriptions of AGN jet acceleration. The three examples cover (1) stability of relativistic jets upon encountering density discontinuities; (2) transverse stability of two-component jets; and (3) propagation characteristics of helically magnetized, high-speed flows.

3.1 Decelerating Jets at Contact Discontinuities

Our first example is motivated by observations, which reveal that in many Faranoff-Riley type I (FR I) radio galaxies, the associated jets are relativistic on parsec scale and sub-relativistic on kiloparsec-scales. This implies that jet deceleration, and thus energy redistributions, must happen within kiloparsec scales [8]. Since the jets travel for enormous distances, it is inevitable that they encounter various sudden transitions of interstellar medium properties. These could be traveling shock fronts, more gradually varying background variations as one traverses regions of differing gravitational potential, or contact discontinuities indicative of boundaries between varying regions of influence. We concentrate on the latter, representing density (and entropy) changes across which total force balance holds (uniform pressure), as these are invariably found in any kind of hydrodynamic interaction involving winds, outflows, etc. In fact, many simulations of AGN jets to date address morphology aspects due to mass entrainment when relativistic jets encounter higher density environments. A decisive parameter in these studies is then the ratio of density contrast assumed between jet beam and "molecular cloud" environment. We here present a study of relativistic hydro jets which undergo sudden deceleration events as they encounter contact discontinuities, as studied extensively in [25]. This generalizes results from Martí et al. [21], who investigated relativistic jets injected in uniform media at fixed Lorentz factor and density contrast. Here, we augment those findings by (1) the use of a more realistic equation of state (a constant polytropic index was assumed in their work) and (2) for the first time addressing how a preshaped jet interacts with a contact discontinuity, forming a new explanation for sudden jet deceleration. The use of a realistic equation of state, where the effective polytropic index $\hat{\gamma}$ in the specific internal energy $p/(\hat{\gamma} - 1)\rho$ varies from 4/3 at relativistic internal energies to 5/3 at classical temperature ranges, is particularly important to quantify this deceleration. This is because its precise value affects the compression rate and the shock strength. In order to follow the jet as it propagates through layers characterized by different background density, the use of AMR for relativistic computations is an absolute requirement.

Our simulation adopts a $\Gamma \approx 20$ beam Lorentz factor and corresponds to an observationally reasonable jet luminosity [5] associated with the jet kinetic energy of about 10^{46} ergs/s. It considers a jet with radius 0.05 parsec at very high relativistic beam Mach number of order 1200, and we follow its course for a distance up to 400 jet radii (i.e., up to 20 parsec). Using AMR, the runs discussed below achieve effective resolutions of 640×19200 or more. The jet traverses half of this distance through lower density surroundings. A contrast of $\rho_{\text{low}}/\rho_{\text{b}} = 0.1496$ is assumed in proper density, so that little deceleration can happen in this lower region. For the mentioned jet beam and density values, the frontal bow shock is then estimated to travel at a relativistic velocity of about Lorentz factor 5. This low density is justifiable, as it could be a region previously cleared by separate outflow events, or simply correspond to typical low density ISM conditions of number density $n = 1\,\text{cm}^{-3}$. A zoom on the jet head, seen in the variation of the effective polytropic index, after 160 radial light crossing times is given in Fig. 10, top panel. It can be seen that the very frontal part consisting of shocked ISM medium is in the relativistic regime with low polytropic index. We investigate its influence on the further dynamics and show in the next two panels the jet head density structure after it encounters a density discontinuity, for a now reversed density contrast where $\rho_{\text{up}}/\rho_{\text{b}} = 4.687$ (middle panel) and one with $\rho_{\text{up}}/\rho_{\text{b}} = 617.22$ (bottom panel). In both cases, the jet survives the density encounter, but a fairly stable beam is found in the first, while a turbulent cocoon enriched by (shear flow unstable) backflows is found in the second. The higher environment density also induces more axial jet confinement. The most important observation is that the first case has its forward shock remaining relativistic, while the second has reduced compression rate at the forward shock, which has now become Newtonian (i.e., with classical 5/3 local polytropic index). As a result, the second case can be quantified to propagate with sub-relativistic velocities in the very dense upper medium, while the first undergoes little deceleration from its initial Lorentz factor 5 propagation. Hence, these simulations serve to show that unexpectedly high-density contrasts need to be used to find significant deceleration.

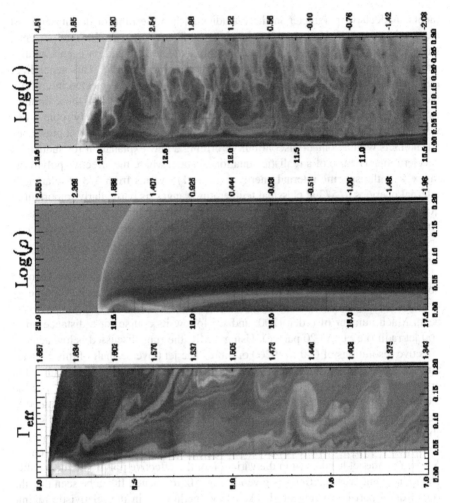

Fig. 10 (Color online) *Top*: The variation in effective polytropic index for a relativistic jet after 160 light travel times through lower density medium, showing the high compression at its bow shock. *Middle*: The density in the jet head after penetrating a higher density cloud of about 5 times higher proper density. *Bottom*: Density variation when penetrating a 600 fold higher density medium. The jet propagates from right to left

3.2 Two-Component Jet Stability

A second example reviews recent findings from [24], where in contrast to the axisymmetric simulations adopted in the previous Sect. 3.1, we now investigate stability when transverse jet structure is at play. In fact, many observational and theoretical arguments are in favor of radially structured jets: a fast inner jet could be the result of a general relativistic launching process acting near the central AGN engine, while a more classical, (magneto-)centrifugal launching effect gives rise to a second, slower outflow component from the surrounding accretion disk. A recent study concentrated on the topological stability in axisymmetric MHD settings for such two-component jets [22]. When looking in a horizontal cross-section through the jet beam at sufficient distance from the source, we then find a two-component jet structure, with a radial interface separating the axially oriented, relativistic outflows with different properties. One can argue for the existence of a fast ($\Gamma \approx 16$) hot inner jet, surrounded by a cold outer jet with reduced Lorentz factor of $\Gamma \simeq 3$. The different launch mechanisms will imprint different rotation profiles as well, and this fairly realistic two-component relativistic jet was again modeled taking the varying polytropic index effects into account. In terms of Mach number, the inner jet is at about 23 and the outer jet is of about Mach 5.

One finds that a body mode develops in the inner jet, which changes from axisymmetric to $m = 4$ azimuthal wave number character, and this mode interacts with a surface Kelvin–Helmholtz mode that originates at the radial interface. Their interaction and nonlinear evolution causes significant compressive disturbances invading the outer cold jet, which expands and this in turn induces Rayleigh–Taylor type instabilities perturbing the outer edge of the cold jet. Still, the overall jet remains coherent and appears collimated in the process. Figure 11 shows the density and effective polytropic index variation near the end of the simulated interval. The outer cold jet shows distinct counter-rotating vortices which are relativistically hot. These could be relevant sites for particle acceleration. Obviously, this result should be revisited in fully three dimensional simulations, where additional axial mode development is allowed. Note though that this two dimensional AMR simulation achieves a 2048^2 grid resolution, and a similarly high axial resolution must be ensured in three dimensions. Moreover, the pure relativistic hydro assumption must be relaxed, since the influence of magnetic fields on the interface Kelvin–Helmholtz

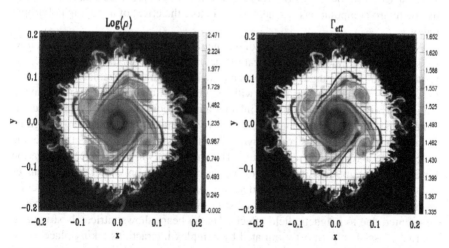

Fig. 11 (Color online) Logarithm of density (*left*) and effective polytropic index (*right*) in a transverse cross-section of a two-component jet. Various instabilities have developed nonlinearly, but an overall stationary configuration remains with a cold outer jet surrounding a fast hot inner jet

development must be properly accounted for, as should be clear from the discussion presented in the first part of this computational survey.

3.3 Relativistic MHD Models with Helical Fields

Our final example of large-scale AGN jet modeling summarizes findings in [17]. There, the relativistic jet is again assumed to be axisymmetric and the propagation characteristics of helically magnetized jets are of interest. Note that due to the axisymmetry assumption, this model does not address dynamics resulting from current-driven kink modes as mentioned in Sect. 2.5, since those involved helical displacements of the jet axis.

The equations used to model these jets are the Lorentz invariant set of ideal relativistic MHD, combining particle, tensorial energy–momentum conservation, and the full complexity of Maxwell's equations, including the displacement current contribution. The ideal MHD assumption boils down to still assume vanishing electric fields in the co-moving frame, i.e., $\mathbf{E} = -\mathbf{v} \times \mathbf{B}$, with all the usual three-vectors expressed in a fixed Lorentzian reference frame. A space-time split (3+1) can be adopted, where the conserved variables include the magnetic field $\mathbf{B}$. Then, a set of conservation laws are obtained which can be combined with the usual AMR methodologies. Due to the extreme contrasts associated with relativistic phenomena, the requirements to ensure numerically a positive proper density, pressure, (partial) energy density, while keeping $\Gamma \geq 1$, and solenoidal magnetic fields are fairly stringent. As a robust second-order scheme essentially using the fastest magnetoacoustic signal speed only, the TVD Lax–Friedrichs method [29] is used, with a diffusive control of the $\nabla \cdot \mathbf{B}$ errors. In contrast to both previous examples of relativistic hydro computations, we here as yet ignore the effect of a varying polytropic index.

Novel to earlier modeling efforts of magnetized relativistic jets [18, 19], we allow for significant variation of the jet beam structure: the Lorentz factor has an average value of about 7, but ranges from 22 on axis to about 4 at the jet radius. The jet has a fully helical internal field structure, with the inverse pitch $\mu = R_{jet} B_{\varphi} / R B_{Z}$ having values of about 0.5 throughout. The inlet parameters consider a kinetic energy dominated, near equipartition (magnetic energy versus internal thermal energy) regime, and similar to the end phase of the hydro simulations mentioned in Sect. 3.1, let the jet penetrate a 10-fold denser medium. The simulations follow the jet propagation for about 150 light crossing times of the jet radius, and only after such distances did the on-axis jet Lorentz factor reduce to values between 5 and 10. The difficulty to decelerate highly relativistic jets thus returns in these simulations and is partly augmented due to magnetic field effects. The jet beam shows intricately structured internal cross-shocks (mostly initiated by complex interactions taking place at the termination shock of the jet beam, with vortex sheddings occurring at the contact interface between shocked jet beam, and shocked ISM matter). Across these continuously reforming shocks, helicity changes are clearly noticeable, as e.g., demon-

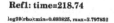

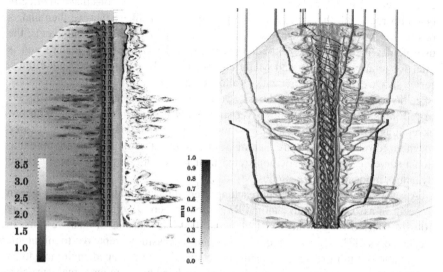

Fig. 12 (Color online) *Left panel*: A greyscale plot of the logarithm of density (*left*) and magnetic pitch μ (*right*) after 147 light crossing times of the jet radius R_{jet}. The light jet is injected from the bottom with average Lorentz factor 7. *Right panel*: The magnetic configuration combined with a translucent impression of the density. The domain shown extends from $[-20, 20] \times [0, 160]$

strated in Fig. 12 at right. In this jet, the inlet (average) Mach number is at about 7. The magnetic field compressions aid to re-accelerate the flow to high velocity, making it harder to decelerate this well-collimated flow. The helicity is then fairly well preserved over significant propagation distances. The flow can be quantified in terms of its advance speed of the bow shock marking the jet head: we find that it is consistent with Lorentz factors above the jet beam average. This relates to the central core of ultra-relativistic flow speeds ($\Gamma \approx 22$) which was assumed at the inlet. Once more, fully three dimensional simulations are called for, in order to address true stability properties over even more extended domains. They will certainly be computationally demanding, as the axisymmetric results reviewed here already correspond to effective resolutions of 3200×8000, over significantly long timescales.

4 Summary and Outlook

In this survey, we touched upon stability aspects of magnetized classical as well as relativistic jets. We highlighted earlier results where either very idealized (Sect. 2) or sufficiently complex (Sect. 3) jet models are employed. Both methods complement each other: the first allows for detailed mode development and interaction effects to be studied in isolation, while all these intrinsically happen in the latter approach. From the observational vantage point, particularly the latter is found to appeal since observed morphologies can be compared, and the outcome of the high-resolution computations can in turn be used to produce synthetic observations that can directly be compared. Still, their increased complexity makes it likely that we then under-resolve various important aspects of astrophysical jet dynamics. In all events, the use of massively parallel, grid-adaptive frameworks is becoming mature and is vital for relativistic hydro and MHD computations. In the employed algorithms, robust shock-capturing discretizations allow to reach very high effective resolutions. Specific future work should address two-component, magnetized relativistic jets in full three dimensional studies, over sufficiently long-term evolutions. With few exceptions [10], quantitative stability studies for more realistic, internally structured magnetized relativistic jets are lacking at this point. The first part of our computational survey convincingly demonstrated their power in interpreting the nonlinear dynamics in idealized case studies.

A severe challenge also lies in reaching even higher Lorentz factor flows known to occur in association with Gamma Ray Bursts. There, Lorentz factors exceeding 100 are inferred for the outflows that signal the violent deaths of massive stars. In recent works [23], AMR techniques have been successfully employed to mimic the deceleration of ultra-relativistic shells, traversing the wind-shaped environments of long GRB progenitor stars. This has yet to be modeled by using more realistic equations of state and/or allowing for strong magnetic fields. Poynting flux-dominated jets can be even more liable to current-driven instabilities than the kinetic energy ones mentioned in Sect. 3.3, so that three dimensional simulations will be required from the outset. The combination of high Lorentz factor flows, dominant magnetic fields, and extreme length and time scale requirements will be a challenge for future computations.

Acknowledgements Computations have been performed on the K.U. Leuven High performance computing cluster VIC. ZM and RK acknowledge financial support from the Netherlands Organization for Scientific Research, NWO grant 614.000.421, and from the FWO, grant G.0277.08.

References

1. Baty, H., 2005, A&A 430, 9
2. Baty, H., Keppens, R., 2002, ApJ 580, 800
3. Baty, H., Keppens, R., Comte, P., 2003, Phys Plasmas 10, 4661
4. Baty, H., Keppens, R., 2006, A&A 447, 9
5. Celotti, A., Ghisellini, G., 2008, MNRAS 325, 283
6. De Sterck, H., Low, B.C., Poedts, S., 1998, Phys. Plasmas 5, 4015
7. De Sterck, H., Low, B.C., Poedts, S., 1999, Phys. Plasmas 6, 954
8. Giovannini, G., 2004, Astrophys Space Sci 293, 1
9. Goedbloed, J.P., Poedts, S., 2004, Principles of Magnetohydrodynamics, Cambridge University Press, Cambridge
10. Hardee, P.E., 2007, ApJ 664, 26
11. Jeong, H., Ryu, D., Jones, T.W., Frank, A., 2000, ApJ 529, 536
12. Jones, T.W., Gaalaas, J.B., Ryu, D., Frank, A., 1997, ApJ 482, 230
13. Kellermann, K.I., Lister, M.L., Homan, D.C., et al., 2004, ApJ 609, 539

14. Keppens, R., Tóth, G., 1999, Phys Plasmas 6, 1461, A&A 491, 321
15. Keppens, R., Tóth, G., Westermann, R.H.J., Goedbloed, J.P., 1999, J Plasma Phys 61, 1
16. Keppens, R., Nool, M., Tóth, G., Goedbloed, J.P., 2003, Comp Phys Commun 153, 317
17. Keppens, R., Meliani, Z., van der Holst, B., Casse, F., 2008, A&A 486, 663
18. Komissarov, S.S., 1999, MNRAS 308, 1069
19. Leismann, T., Antón, L., Aloy, M.A., Müller, E., Martí, J.M., Miralles, J.A., Inabnez, J.M., 2005, A&A 436, 503
20. Malagoli, A., Bodo, G., Rosner, R., 1996, ApJ 456, 708
21. Martí, J.M., Müller, E., Font, J.A., Ibáñez, J.M.A., Marquina, A., 1997, ApJ 479, 151
22. Matsakos, T., Tsinganos, K., Vlahakis, N., et al., 2008, A&A 477, 521
23. Meliani, Z., Keppens, R., 2007, A&A 467 L41
24. Meliani, Z., Keppens, R., 2007, A&A 475, 785
25. Meliani, Z., Keppens, R., Giacomazzo, B., 2008, A&A 491, 321
26. Miura, A., Pritchett, P.L., 1982, JGR 87, 7431
27. Palotti, M.L., Heitsch, F., Zweibel, E.G., Huang, Y.-M., 2008, ApJ 678, 234
28. Ryu, D., Jones, T.W., Frank, A., 2000, ApJ 545, 475
29. Tóth, G., Odstrčil, D., 1996, JCP 128, 82
30. van der Holst, B., Keppens, R., 2007, JCP 226, 925
31. Zaliznyak, Yu., Keppens, R., Goedbloed, J.P., 2003, Phys Plasmas 10, 4478

Jets and Outflows from Collapsing Objects

Robi Banerjee

Abstract Jets and outflows are ubiquitously observed around young stellar objects. There is now strong evidence that these jets are launched from the protostellar disc around the young stars through the coupling of magnetic fields. Magnetic fields threading the pre-stellar molecular cores are dragged inwards during the gravitational collapse and are wound up by the rotating gas in the protostellar disc. The resulting geometry of the magnetic field is that of a hourglass. The magnetic flux is strongly compressed inside the central region and flux lines pointing outwards connecting to the outer region. Additionally, the magnetic field lines anchored to the underlying protodisc are wound up and acquire a strong toroidal component. Such a field configuration, together with the underlying rotor, is known to launch and accelerate material off the disc. This could be the onset of the observed jets around young stellar objects.

In this contribution to the Jetset lecture notes we summarise the research progress in the field of jet launching from collapsing objects. Complying with this workshop on high-performance computing in astrophysics this is done while focusing on results from numerical simulations for this task.

1 Introduction

One of most significant manifestations during the early phase of star formation is the appearance of jets and outflows (see e.g. [6, 3, 15, 7, 4]). It now becomes evident that these phenomena are connected to the birth of both low- and high-mass stars. Low mass stars, the well-known Herbig-Haro objects, are accompanied by highly collimated optical jets (see, e.g. [22, 71]) whereas high-mass stars are often obscured, hard to observe, and much less collimated (e.g. [74, 73, 15]). Nevertheless, there is strong evidence that the underlying launching process is based on the same physical mechanism, namely the magneto-rotational coupling: magnetic fields anchored to an underlying rotor (e.g. an accretion disc) will carry along gas which will be

R. Banerjee (✉)

Institute for Theoretical Astrophysics at the Zentrum für Astronomie of the University Heidelberg, Albert-Ueberle-Str. 2, Heidelberg, Germany, banerjee@ita.uni-heidelberg.de

Banerjee, R.: *Jets and Outflows from Collapsing Objects.* Lect. Notes Phys. **791**, 201–222 (2009)
DOI 10.1007/978-3-642-03370-4_8 © Springer-Verlag Berlin Heidelberg 2009

flung outwards (the same mechanism could also apply to galactic jets, e.g. [69, 70]). With the seminal theoretical work by Blandford and Payne [16] and Pudritz and Norman [68] the idea of magneto-centrifugally driven jets was first established. Herein, it was shown that magnetic fields anchored to the disc around a central object can lift off gas from the disc surface. A magneto-centrifugally driven jet will be launched if the poloidal component of magnetic field is inclined with respect to the rotation axis with an angle of more than 60°. Numerical simulations of Keplerian accretion discs threaded by such a magnetic field then showed that these jets will be self-collimated and accelerated to high velocities (e.g. [31, 61]). This driving mechanism, where the launching of the jet is connected to the underlying accretion disc, predict that jets rotate and carry angular momentum off the disc. Recent HST observations of the optical jet from DG Tauri confirmed these predictions of rotating outflows [5].

In the presence of magnetic fields gas is accelerated by the Lorentz force. Generally, this force can be divided into a magnetic tension and a magnetic pressure term. In an axi-symmetric setup, the magnetic tension term is responsible for the magneto-centrifugal acceleration and the jet collimation via the hoop stress. Also the magnetic pressure can accelerate gas off the underlying disc. These magnetic pressure driven winds are known as *magnetic twist* [75], *plasma gun* [24], or *magnetic tower* [47].

Protostellar discs around young stellar objects themselves are the result of gravitational collapse of molecular cores. Additionally, the interstellar medium and molecular clouds are permeated by magnetic fields of varying strength and morphology (see e.g. [26, 12]). Therefore, there should be a profound link between the collapse and magneto-rotationally driven outflows.

It is a difficult task studying jets and outflows during the collapsing phase of the protostellar evolution. It is observationally challenging because these regions are highly obscured and the star-forming region is not yet cleared from the surrounding medium. Unless the protstellar jet is penetrating such a dense environment it will be difficult to observe. But also theoretical difficulties prevent a rapid progress in this field. The coupling of magnetic field with a self-gravitating system is a highly nonlinear problem where linear simplifications are not sufficient to describe such a configuration. Even nonlinear self-similar solutions do not nearly capture the complexity of such a system. One promising approach to study the self-consistent jet launching during the collapse of self-gravitating gas is to use direct numerical simulations. Even then, the large dynamical range of length scales (from the 10^4 AU molecular core to the sub-AU protostellar disc) and time scales (from the initial free fall time of 10^5 years to the orbital time of one year) require expensive adaptive mesh refinement (AMR) or SPH simulations which can handle magnetic fields. Furthermore, non-ideal coupling effects like Ohmic dissipation and ambipolar diffusion complicate such calculations. One of the first collapse simulations in which outflows are observed were done 10 years ago by Tomisaka [78] with an axi-symmetric nested grid technique. These simulations of magnetised, rotating, cylindrical cloud cores showed that a strong toroidal field component will be build up which eventually drives bipolar outflows. Ever since a number of collapse simulations from different groups and different levels of sophistication where performed (among these are work by [79, 19, 55, 37, 48, 50, 49, 87, 51, 8, 9, 53, 54, 36]).

2 Numerical Modelling

This section describes briefly the different numerical methods and models used to study outflows in collapse simulations.

2.1 Methods and Models

As mentioned earlier, one of the first simulations combining a self-gravitational collapse with magneto-hydrodynamics (MHD) in which outflows where observed was done by [78]. These simulations are based on a *nested-grid* technique in which the simulation box is covered by fixed, overlapping grids of increasing resolutions. The idea to use nested resolutions in numerical methods was initially developed by Berger and Oliger [14] and Berger and Colella [13], and later implemented, for instance, by [72] and Yorke et al. [85] to study self-gravitating systems. In general, physical systems for which useful study one needs a large dynamical range, like collapsing self-gravitating objects, require a special numerical treatment of the spacial structure of the underlying mesh or a Lagrangian treatment of the fluid as provided by the *smoothed particle hydrodynamics* (SPH) approach. In particular, for self-gravitating systems both methods need a minimum resolution to avoid spurious, i.e. numerical, fragmentation. Grid-based codes must obey the so-called Truelove criterion which requires that the *local* Jeans length is always resolved by at least four grid cells [80]. A similar criterion applies for particle-based codes where the local Jeans mass sets the upper limit for SPH particles [11].

Besides the number of proprietary numerical tools which are suitable for magneto-hydrodynamical collapse calculations, there are also a number of publicly available grid-based codes. Among those, the most widely used are FLASH [35], ENZO [60], NIRVANA [87], RAMSES [34], PLUTO [57], and ATHENA [77]. All these codes are based on an adaptive mesh refinement (AMR) technique and support MHD (for the MHD implementation to the ENZO code see [23, 83, 84]). On the side of SPH, there are several developments to implement MHD in the Lagrangian treatment, but so far no version is publicly available (e.g. [56, 38, 37, 18, 86]). Difficulties to handle magnetic fields in the SPH approach arise due to erroneous nondivergence-free components of the magnetic field. A very promising solution to this problem based on a hyperbolic divergence cleaning method is outlined and extensively tested by [62–64].

The study of outflows from collapsing objects with numerical simulations includes a variety of different setups, from cylindrical cores in axi-symmetric calculations (e.g. [78, 79]) to fully three-dimensional calculations of collapsing spheres (e.g. [55, 52, 8, 36]), or even shearing box simulations in which outflows are launched

from giant planets [51]. In what follows, we will focus on the collapse of magnetised, rotating Bonnor–Ebert spheres based on three-dimensional simulations done
with the FLASH code.

2.2 An Example Setup

As an initial setup we start with a rotating, self-gravitating Bonnor–Ebert sphere
which is embedded in a magnetised low-density medium (see 8, for details). Bonnor-
Ebert spheres ([30, 17], BE-sphere) are isothermal gas cores which are in hydrostatic
equilibrium. Similar to a overdensity which becomes Jeans unstable, there exists a
critical radius, if exceeded, the BE-sphere will collapse under its own weight. In
units of $r_{BE} = c/\sqrt{4\pi\,G\rho_c}$ (where c is the speed of sound, G the gravitational
constant, and ρ_c the core density) this radius ξ has the value of 6.451. Nowadays,
there are many observations of molecular cores which can be perfectly fitted with
such a BE-profile (e.g. [42]). The most prominent one is the Bok globule, Barnard
68, studied by Alves et al. [2] with extinction measurements. We use these measured
values for our initial setup: core density $\rho_0 = 9.81 \times 10^{-19}\,\mathrm{g\,cm^{-3}}$, mass of the
cloud core $M = 2.1\,M_\odot$, radius $R = 1.25 \times 10^4\,\mathrm{AU}$ ($\xi = 6.9$), and overall gas
temperature $T = 16\,\mathrm{K}$. On top of this static cloud we assume a solid-body rotation
of the cloud core with an angular velocity of $\Omega = 1.89 \times 10^{-13}\,\mathrm{rad\,s^{-1}}$ which
corresponds to $\Omega\,t_{ff} = 0.4$, where $t_{ff} = 2.12 \times 10^{12}\,\mathrm{s}$ is the initial free-fall time.
With these parameters the rotational energy is a few percent of the gravitational
energy of the sphere (see Fig. 1 for an illustration of the initial setup).

The initial magnetic field is setup to be parallel to the rotation axis (z-axis). To
account for a magnetic pressure enhancement during core formation we assume a
constant thermal-to-magnetic pressure, $\beta = p/(B^2/8\pi)$, in the equatorial plane
with a value of $\beta = 84$ which gives a minimal and maximal field strength of $B_{min} =
3.4\,\mu\mathrm{Gauss}$ and $B_{max} = 14\,\mu\mathrm{Gauss}$, respectively. Note that this field configuration
is not force-free (i.e. $(\nabla \times \mathbf{B}) \times \mathbf{B} \neq 0$) due to the variation of B_z in the x–y
plane. Nevertheless, this force is small compared to the force resulting from the
thermal pressure. The field is also weak with respect to the self-gravitating mass and
therefore initially dynamically unimportant. The mass-to-flux ratio of this system
exceeds the critical value $0.12/\sqrt{G}$ by more than a factor of 10.

Before we proceed with the discussion of the results from these simulations we
like to stress that these calculations are based on the assumption of *ideal* MHD.
Although, it is unlikely that the general results obtained from ideal MHD calculations are invalid, quantitative details could change if non-ideal effects are included.
For instance, Machida et al. [53] showed that jets and outflows will be launched
during the protostellar phase even if Ohmic dissipation of magnetic flux might
become important. More crucial could be the influence of *ambipolar diffusion*,
where the magnetic field lines start to slip relative to the neutral gas in the weakly
coupled ion-neutral regime. In principle, less magnetic flux could be trapped in the
high-density regions if ambipolar diffusion becomes important which in turn could

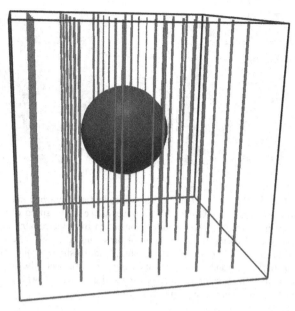

Fig. 1 (Color online) Shows an example setup in which a rotating Bonnor–Ebert sphere is embedded in a weakly magnetised, low-density environment (see text for details). The subsequent collapse and spin-up of the inner core generates toroidal fields which are strong enough to launch outflows from the protostellar disc (see Figs. 3 and 5)

reduce the influence of magnetic fields in these regions. But even then, recent results from collapse calculations by Duffin and Pudritz [29], where the effect of ambipolar diffusion is taken into account, confirmed the general picture summarised in what follows.

3 Magnetic Braking in the Pre-collapse Phase

A spinning, magnetised core undergoes significant magnetic braking even before the collapse begins. In this early phase, some of the core's initial angular momentum will be carried off by a flux of torsional (Alfvénic) waves which transfer angular momentum from the spinning sphere to the surrounding gas gradually spinning down the sphere. The initial magnetic braking of a spinning cloud core was studied analytically by Mouschovias and Paleologou [59] for idealised disc-like rotors embedded in a non-rotating environment, and its appearance provides an excellent test of the accuracy of our numerical methods as well. One can compare such a situation with a rotating object linked with wires to non-rotating surroundings. Due to the connection through magnetic field lines (which are the wires in this picture) the non-rotating environment wants to co-rotate with the rotating object. Therefore, angular momentum is extracted from the underlying disc or sphere while the external medium starts to rotate.

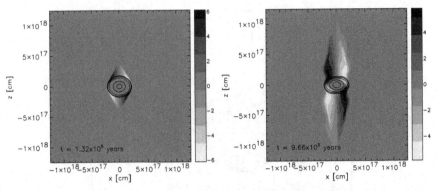

Fig. 2 (Color online) Shows the torsional Alfvén waves that are launched into the non-rotating ambient medium (color scale in μGauss). These waves move at the speed of the local Alfvén velocity ($v_A = 486\,\mathrm{m\,s^{-1}}$) and transfer angular momentum from the rotating cloud core to the low-density environment leading to a slow down of the cloud core [59]. The snapshots which show the toroidal magnetic field (gray scale in μGauss) and density (*contour lines*) are taken at $t = 1.3 \times 10^5$ years (*left panel*) and $t \sim 1 \times 10^6$ years (*right panel*), respectively. The density *contour lines* show the location of the Bonnor–Ebert sphere in the simulation box, where the side length of simulation box is 2.5×10^{18} cm $= 1.67 \times 10^5$ AU) (from [8])

In Fig. 2 one can see how the initially stationary, external medium starts to rotate while a torsional Alfvén wave is propagating outward. The subsequent spin down time of a homogeneous disc embedded in a non-rotating environment with a density contrast $\delta = \rho_{\mathrm{cloud}}/\rho_{\mathrm{ext}}$ can be estimated by [59]:

$$\tau_{\mathrm{damp}} \approx \frac{Z\,\delta}{v_{\mathrm{A,ext}}}, \tag{1}$$

where $v_{\mathrm{A,ext}} = B/\sqrt{4\pi\,\rho_{\mathrm{ext}}}$ is the Alfvén velocity in external medium and Z is the disc height. For our setup, this gives a spin-down time of $\sim 10^6$ years, which is in good agreement with our numerical results.

In order to generate sufficiently strong toroidal magnetic fields to get magnetically driven outflows, the spin-down time must not be too small. Our simulations with stronger magnetic fields and/or slower initial rotation showed no, or only a weak outflow as the build-up toroidal magnetic field component is not strong enough to power a wind. By comparing the damping time of Eq. (1) to the initial orbital time of the sphere one can estimate whether magnetic braking spins down the sphere too quickly. The condition that the spin-down time not be too short gives (cf. Eq. (18a) in [59]):

$$\tau_{\mathrm{damp}} > 2\pi\,\Omega^{-1}. \tag{2}$$

Using $Z \approx c_s/\sqrt{4\pi\,G\,\rho_0}$ this relation can be rewritten as

$$\frac{(\delta \beta)^{1/2}}{\sqrt{3}\,\pi}\, t_{\text{ff}}\, \Omega > 1 \tag{3}$$

(in our case $t_{\text{ff}}\, \Omega\, \sqrt{\delta \beta / 3}/\pi = 2.13$). Simulations where condition Eq. (3) is not or only barely fulfilled result in a more spherical collapse with less pronounced discs and had no tendency to fragment. We conclude from these results that modest to strong magnetic fields stabilise the cloud core and prevent fragmentation during the collapse. A measure of the toroidal field under these circumstances is given by the expression ([59], Eq. (17a))

$$\frac{B_\phi}{B_z} = \frac{R\,\Omega}{v_{\text{A,ext}}}, \tag{4}$$

where B_z is the initial, homogeneous, magnetic field. For our initial setup B_ϕ / B_z becomes larger than one at a disc radius of 1.14×10^{17} cm.

The theory of magnetic braking also predicts that the deceleration of the disc is more efficient if the magnetic field is parallel to the disc plane (i.e. perpendicular to the rotation axis [58]). Recent simulations of collapsing, magnetised Bonnor–Ebert spheres [55] showed that the rotation axis will be aligned with the magnetic field during the collapse of the cloud core as the perpendicular (to the magnetic field) component of the angular momentum is extracted more quickly than the parallel component. Therefore, an aligned rotator might be a "natural" configuration of magnetised cloud cores.

4 The Initial Collapse Phase

The first phase of the collapse of magnetised Bonnor–Ebert spheres proceeds very similarly to the non-magnetised cases (see, e.g. [43, 33, 10, 8, 54]): initially the rotating cloud core collapses from outside-in. As long as the core is optically thin to infrared dust emission, the collapse proceeds isothermally. Only when the core density exceeds $\sim 10^{10}$ cm^{-3} the infalling material falling becomes affected by the magnetic field. Even at this stage, the magnetic pressure is still smaller than the thermal pressure ($\beta > 1$).

Note that a general prediction of the density at which the outflow will be launched will involve the knowledge of the magnetic field structure and strength as well as the rotation velocity of the first core. The non-linear interplay between the magnetic field and the self-gravitational collapse of a rotating cloud hampers this task. Nevertheless, we find that the outflow starts well into the collapse and after the first core has formed.

5 Onset of Large Scale Outflow

A strong toroidal magnetic field component builds up by winding magnetic field lines as the core's angular velocity increases during its contraction phase. By the time the central density reaches $\sim 10^{10}\,\mathrm{cm}^{-3}$, the magnetic pressure from the toroidal field component has become strong enough to prevent the shock fronts below and above the disc plane from moving towards the centre. Now, material inside the magnetised bubble is pushed outward leading to a large scale outflow. The onset of this large scale magnetic tower outflow and the situation at the end of our simulation are shown in the two snapshots in Fig. 3. Here, the out-flowing gas has a torus like structure around the rotation axis (see also Fig. 6). This is a result of the magnetic field configuration wound up on large scales during the collapse of a rotating core. In particular, the confinement due to the shock fronts compress the the toroidal field. This results in an increasing magnetic pressure force in the radial direction away from the rotation axis which in turn leads to a torus-like gas outflow structure. Such large scale, low velocity outflows are typical phenomena of rotating, collapsing objects (e.g. [78, 79, 1, 48, 8, 54]).

This collimated bipolar outflow can be understood in terms of a magnetic tower [24, 47, 40] that consists of an annulus of highly wound magnetic field lines that pushes into the ambient pressure environment. The toroidal magnetic field component that is continuously produced by the rotating disc acts like a compressed spring which lifts some material off the disc surface and sweeps up the material in the external medium. Substantial pressure is needed to trap the toroidal field, allowing it to wind up and push into this region. Solutions show [47] that the expansion of a tower grows linearly with time in the case of an uniform external pressure and accelerates in the case of a decreasing pressure profile. The acceleration of a tower may be slowed or even reversed in the presence of ram pressure. Taken together, it appears natural that a tower flow be driven from a disc within the high-pressure region inside the first shock.

At this stage where the large scale outflow begins to sweep up material the magnetic pressure becomes stronger than the thermal pressure ($\beta \sim 0.1 - 1$) and the interior of the "magnetic bubble" cools further by adiabatic expansion. Such an outflow may be the origin of the molecular flow that is seen from all young stellar objects (YSOs) (e.g. [81]). Measurements of CO emission lines of outflows from young stellar objects indicate higher velocities, but our simulations show the star formation phase in a very early stage wherein the central mass of the protostar is still tiny. Since the outflow velocity is related to the escape speed (see [67], for a review), this is the expected result. The outflow speed will increase with time as central stellar mass grows. In Fig. 3 one can see that the outflow velocity exceeds the poloidal Alfvén velocity where the outflow is the fastest. The outflow forces the region enclosed by the outer shock fronts to expand and the shock fronts are moving outward. By the end of the presented simulation the shock fronts are pushed to a disc height of $\sim 600\,\mathrm{AU}$ and would presumably continue to rise. As already mentioned, this may be the origin of bipolar outflows that are associated with all young stellar objects.

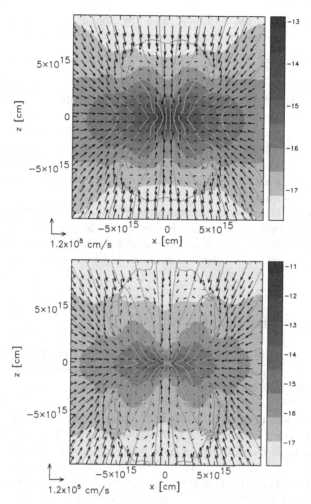

Fig. 3 (Color online) onset of the large scale outflow.The snapshots show the time evolution of the magnetically driven winds which are launched during the star formation process. The *upper panel* shows the collapsing stage at $t = 6.81 \times 10^4$ years shortly after the onset of the outflow and the *lower panel* shows the situation 1430 years later, at the end of our simulation, when the outflow is clearly visible. The magnetic pressure drives a "bubble" which is surrounded by shock fronts and reverses the gas flow. Here, the typical outflow velocity is $v_z \sim 0.4\,\mathrm{km\,s^{-1}}$ at $z = 400\,\mathrm{AU}$. The colour scale shows the gas density distribution (logarithmic scale in $\mathrm{g\,cm^{-3}}$), the vector field reflects the velocity flow, the blue contour lines mark the Alfvén surfaces, and the magnetic flux surface ((B_z, B_x) components) is drawn in green (from [8])

We point out that the collimation of the large scale outflow is not due to the initially uniform field which extends to infinity but rather to the dynamically built up field structure which provides hoop stresses to confine the outflow. The fact that toroidal field component dominates the poloidal component in the outflow region shows that the large-scale outflow is driven by (toroidal) magnetic pressure and confined by the same toroidal field structure as shown in Lynden-Bell [47].

6 Outflows from Massive Star-Forming Regions

Outflows are not only connected to low mass objects. They are also frequently observed around young massive stars and are believed to be as common as outflows from low mass YSOs (see e.g. [15, 4]). Numerical calculations of collapsing massive cloud cores also predict that outflows will be launched from the protostellar disc if the ISM is magnetised [9].

Similar to the low-mass case, a strong toroidal field will be build up during the collapse of the rotating massive core. Eventually this field will be strong enough to lift off gas from the young massive disc (see Fig. 4). In general, the mechanism of disc-driven outflows are launched and collimated by Lorentz forces (see e.g. [70]).

7 The Onset of the Disc Jet

An even more dramatic outflow phenomenon erupts from the interior regions of the disc, in the deepest part of the gravitational potential well generated by the assembling protostar. In Fig. 5, we show two snapshots of the disc and surrounding infalling region focused down to a scale of $\sim 0.7\,\mathrm{AU}$. The upper panel shows the collapse of material that is still raining down onto the disc at time of $t = 6.8 \times 10^4$ years. In comparison with the outer regions of the disc, the magnetic field lines towards the disc interior have been significantly distorted as they are dragged inwards by the disc's accretion flow. They take the appearance of a highly pinched-in hourglass. This configuration is known to be highly conducive to the launch of disc winds [16, 68, 46, 32]: magnetic field lines threading the disc with an angle with the vertical that is greater than 30° are able to launch a centrifugally driven outflow of gas from the disc surface. Our simulations clearly confirm this picture as the angles of the magnetic field lines with the vertical axis are much greater than 30°.

Five months later in our simulation, a jet can clearly be seen to leave the disc surface inside a spatial scale of 3×10^{12} cm. Moreover, this disc wind achieves super-Alfvénic velocities above which it begins to collimate towards the outflow axis. This jet is much more energetic than the magnetic tower outflow at 1000 times larger scales (cf. Fig. 3).

Similar to the large scale outflow, this jet is confined between shock fronts. The appearance of these shocks seems to encourage flow reversal, possibly because the high pressure inside the post-shock region. The simulations by Machida et al. [54], in which outflows from the first and second core are investigated, show very similar structures of the large scale outflow and small scale driven high velocity jet.

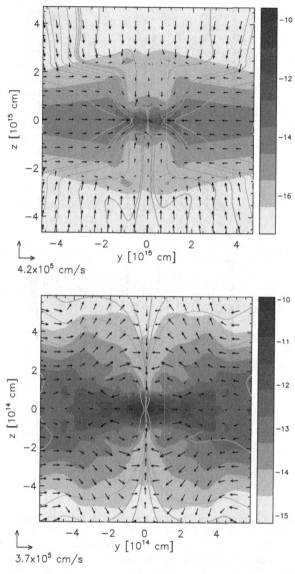

Fig. 4 (Color online) Onset of outflows from collapsing massive cloud cores. Similar to the low-mass case, an outflow will be launched from the protostellar region if the parent cloud is magnetised. Here the results form the collapse of a 170 $M_\odot$ core are shown. Initially the core is modelled with a density distribution following a Bonnor–Ebert profile with a core density of 3.35×10^{21} g cm^{-3} and a radius of 1.6 pc (from [9])

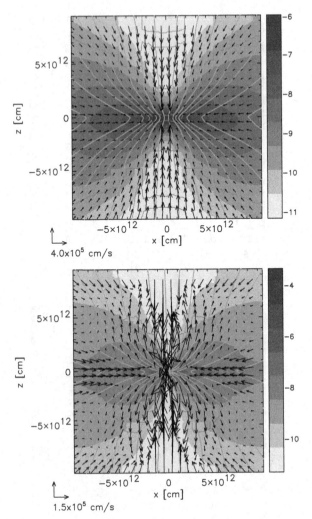

Fig. 5 (Color online) Onset of the jet. The snapshots show the inner structure of the protostar and pre-stellar disc up to $\sim 0.7\,\text{AU}$ (zoom factor 1000 compared to Fig. 3). Only 5 months (1.258×10^7 s) after the upper snapshot is taken (at $t = 6.89 \times 10^4$ years) jets above and below the pre-stellar disc are centrifugally driven by the strong magnetic field. The jet velocities reach $\sim 3\,\text{km s}^{-1}$ at 0.4 AU (from [8])

    Typical self-consistent field configurations from a collapsing cloud core can be seen in Fig. 6, which are results from the numerical simulations described above. The upper panel of this figure shows the three dimensional field structure of magnetic field lines on the scale of the large scale outflow in Fig. 3, while the lower panel shows the same for the jet in Fig. 5. The magnetic field lines in both snapshots are always swept backwards in a rotating outflow – an affect that arises from the mass loading of these flows. Nevertheless, there is a difference in the field line structure – those in the upper panel are almost parallel to the outflow axis outside

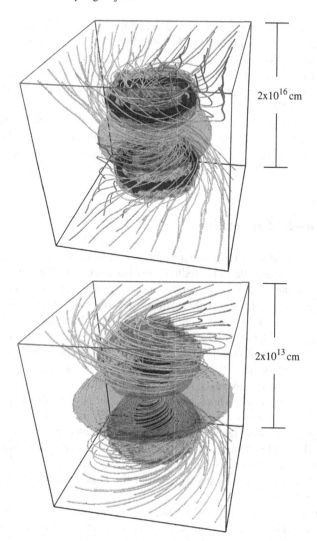

2×10^{16} cm

2×10^{13} cm

Fig. 6 (Color online) magnetic field line structure, outflow, and disc. The two three dimensional images show the magnetic field lines, isosurfaces of the outflow velocities, and isosurfaces of the disc structure at the end of our fully three-dimensional simulation ($t \simeq 7 \times 10^4$ years) at two different scales: the *upper images* refers to the scale shown in Fig. 3 (large scale outflow, side length of the box $L = 1.95 \times 10^{16}$ cm) and the *lower image shows* the jet launching region of Fig. 5 ($L = 1.91 \times 10^{13}$ cm). The isosurfaces of the upper panel refer to velocities $0.18 \, \mathrm{km \, s^{-1}}$ (*light red*) and $0.34 \, \mathrm{km \, s^{-1}}$ (*red*) and a density of $2 \times 10^{-16} \, \mathrm{g \, cm^{-3}}$ (*gray*) whereas the lower panel shows the isosurfaces with velocities $0.6 \, \mathrm{km \, s^{-1}}$ (*light red*) and $2 \, \mathrm{km \, s^{-1}}$ (*red*) and the density at $5.4 \times 10^{-9} \, \mathrm{g \, cm^{-3}}$ (*gray*) (from [8])

the large scale outflow region, while the field lines in the jet region have a strong component parallel to the protodisc. The latter field structure indicates that the jet is magneto-centrifugally powered wherein the field lines act as "lever arms" flinging material off the disc surface, whereas the large scale outflow is powered by magnetic pressure in which the magnetic field lines behave like the release of a compressed spring (see also [25]).

Very few observations of magnetic field line structures around young stellar objects do exist up today. But recent Zeeman measurements of the FU Ori disc reveal field line structure reminiscence of the configuration seen in collapse simulations [28].

8 The Disc–Jet Connection

Outflows and jets from young stellar objects are powered by the accretion flows in the disc where magnetic fields redirect a small fraction of the flow leading to a mass loss from the protodisc. Therefore, accretion properties and wind properties should be strongly correlated. Observations of various different systems indicate for instance that the mass loss by the wind $\dot{M}_{\mathrm{wind}}$ and the mass accretion through the disc $\dot{M}_{\mathrm{accr}}$ are related by $\dot{M}_{\mathrm{wind}}/\dot{M}_{\mathrm{accr}} \sim 0.1$ (e.g. review [41]). In Fig. 7, we summarise the results of our simulation which compares the wind properties with that from the accretion disc 70, 000 years after the onset of the gravitational collapse (quantities are shown as averaged over a sphere with radius r). Clearly visible in these graphs are the double-shock structures that confine the jet and the large scale outflow to distinct regions. The inner shock (~ 1 AU) momentarily hinders the jet from expanding towards higher disc altitudes and the outer shock (~ 600 AU) until now blocks the outflow from running into the cloud envelope. Nevertheless, the jet is already powerful enough to expel material at a rate of $\dot{M}_{\mathrm{jet}} \simeq 10^{-3} M_\odot/$year at $r = 3 \times 10^{12}$ cm which is a good fraction of the peak accretion through the disc ($\dot{M}_{\mathrm{jet}}/\dot{M}_{\mathrm{accr}} \sim 1/3$).

The system has not yet reached a steady state configuration and is still in a contracting phase. Therefore, the large scale outflow is not yet fully developed and its contribution to the mass loss and luminosity is still gaining importance. Nonetheless, we can classify the last stage of our simulation as a "Class 0" protostar because the luminosity of the large scale outflow might be in the observable range (a few % of the accretion luminosity).

Even at this early times the small scale jet already carries a significant amount of the total angular momentum which is extracted from the protodisc. This can be seen in Fig. 7 which shows the angular momentum per unit length,

$$\frac{dJ^\pm}{dr} = r^2 \int d\Omega \, \rho \, R \, v_\phi, \tag{5}$$

and the total angular momentum

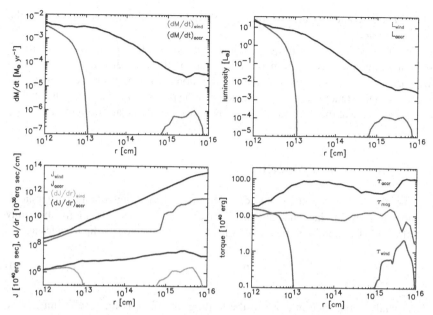

Fig. 7 (Color online) Disc wind connection. These panels show the properties of the disc-wind connection at the end of our simulation, i.e. $t = 7 \times 10^4$ years. From *top left* to *bottom right*: mass outflow and mass accretion through the disc plane; kinetic luminosities from the wind material and gas "raining" onto the disc; angular momenta, J, carried by the wind and by the disc; total torques (change of angular momenta) of the magnetic field, the wind, and accretion disc. The double-shock structure at $r \sim 1$ AU and $r \sim 600$ AU is clearly visible in the wind properties: the jet and the large-scale outflow are still confined beneath these shock fronts. Both the jet and the large scale outflow are launched at different disc heights by different driving mechanisms (from [8])

$$J^{\pm} = \int dr \, \frac{dJ^{\pm}}{dr}, \qquad (6)$$

carried by the infalling (accretion, $-$) and outgoing (wind, $+$) material, where v_{ϕ} is the toroidal velocity and R is the cylindrical radius. The lower right panel of Fig. 7 shows the torques exerted by the accretion flow, wind, and magnetic field, where we calculate the accretion torque ($-$) and the wind torques ($+$) as the angular momentum flux through a sphere with radius r,

$$\tau^{\pm} = r^2 \int d\Omega \, v_r^{\pm} \, \rho \, R \, v_{\phi}, \qquad (7)$$

where $v_r^{\pm}$ is the radial outflow/inflow velocity. The torque exerted by the magnetic field on the fluid at radius r is (e.g. [41])

$$\tau_{\text{mag}} = \frac{1}{4\pi} r^2 \int d\Omega \, (B_r + B_p) \, R \, B_{\phi}, \qquad (8)$$

where B_r, B_p, and B_ϕ are the radial, poloidal, and toroidal components of the magnetic field.

The distribution of the torques indicate that the jet transports almost as much angular momentum away from the inner part of the protodisc as the disc gains from accreting gas. Together with the angular momentum extracted by the magnetic torque from this inner region, we conclude that the protostar(s) will spin much below its break-up rate at the time the system relaxes to a steady-state configuration.

9 Magnetic Field Properties

The distribution of magnetic fields across the disc is a vital ingredient in understanding how outflows are launched and how the bulk of a disc's angular momentum may be extracted. The dominant field component is the vertical one, which scales as

$$B_z \propto R^{-4/3} . \tag{9}$$

This radial dependence is close to the scaling one obtains for a self-similar disc model wherein $B_z \propto R^{-5/4}$ [16]. This is remarkable, however, as the self-consistent collapse is *not* self-similar. This can also be seen by the radial dependence of the toroidal field component which falls off much more steeply than that of the vertical field. We find the scaling close to $B_\phi \propto R^{-2}$ (see Fig. 8). The strength of the dominant field component at 1 AU is quite significant; $B_z(1\text{AU}) \simeq 3.2$ Gauss. It is remarkable that meteorites found at about this physical scale in the solar system are found to have been magnetised by a field strength that is of this order – 3 Gauss [44].

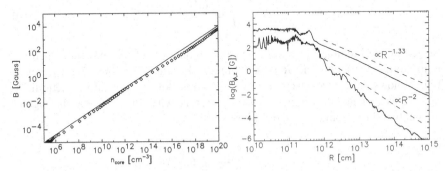

Fig. 8 Shows the magnetic field evolution as a function of the central density (*left panel*) and the distribution of the poloidal and toroidal field components (*right panel*). The $B - n$ relation shows a power law behavior with an index of $\gamma \sim 0.6$ ($B \propto \rho^\gamma$, solid line). This is slightly steeper than what is found in similar simulations without rotation [27] and core formation simulations [45] which found $\gamma \approx 1/2$. The toroidal component of the magnetic field, B_ϕ, is smaller than the poloidal component, B_z, throughout the disc and follows a different power law. Note that the magnetic field trapped in the protostellar cores reaches fields strength of several kGauss (from [8])

The vertical magnetic field in these data levels off in the innermost region as is seen in this panel. The peak field observed is of the order of 3 kGauss. This result is within the limits of measured mean surface magnetic field strengths of stars, which are observed to have values $\simeq$ 2 kGauss [39]. The interesting conclusion is that the magnetic fields in protostars may be fossils of this early star formation epoch in which the magnetic field of the parental magnetised core was compressed into the innermost regions of the accretion discs. Recently it was shown by Braithwaite and Spruit [20] that fossil fields with non-trivial configuration can survive over a star's lifetime.

Recent measurements by Donati et al. [28] of magnetic fields in the accretion disc around FU Ori support many of the theoretical calculations reported here. Firstly, the reported magnetic fields strength is around 1 kGauss at a distance of 0.05 AU from the protostar. Secondly, the inferred field configuration has the structure of a wound-up and compressed magnetic field which azimuthal component points in the opposite direction of the disc rotation and where the poloidal component dominates the field. Thirdly, the magnetic plasma rotates sub-Keplerian which indicates that strong magnetic braking must be taking place.

Furthermore, one expects that the magnetic flux increases during the collapse with increasing density. Typically, one finds a power law relation between the core density and the magnetic field strength. Observations suggest that a power law index around 0.5 (e.g. [26]). Our particular simulation shows a scaling relation of $B \propto n^{0.6}$ (see Fig. 8). This power law is slightly steeper than $B \propto n^{1/2}$ found in similar simulations of collapsing cores (e.g. [27, 45]). In general, the scaling law, $B \propto n^\gamma$, depends on the field geometry. For instance, small scale tangled magnetic fields scale as $B \propto n^{2/3}$. In our case, we have poloidal *and* toroidal field components which lead to a slightly steeper compression of the field strength with density.

The typical geometry of such a rotationally build-up magnetic field is shown in Fig. 9. The field lines are dragged towards the centre of the collapse, where the poloidal component follows the shape of an hourglass. The toroidal component has a symmetric butterfly shape: it has to vanish at the rotation axis and in the disc plane.

The origin of the magnetic fields in accretion discs remains one of the most important, unresolved issues of star formation. Background fields, which are observed on all scales, will be advected inwards during the collapse of the magnetised core on larger scales – as we see in our simulations. It is also theoretically possible for magnetic fields to be generated within discs by dynamo action (e.g. [66, 65, 76, 21, 82]). The simulations of von Rekowski et al. [82] explicitly include a dynamo generation term ("α") in the induction equation. Their results show that dynamo action in a turbulent disc can generate fields as well as disc winds without the need for larger scale external fields. In most of the numerical simulations however, the time span is not sufficient to observe dynamo-generated disc fields as one can follow only a few disc rotations in this collapse situation. More detailed numerical experiments are needed to quantify the importance of dynamo or stellar generated fields to power outflows.

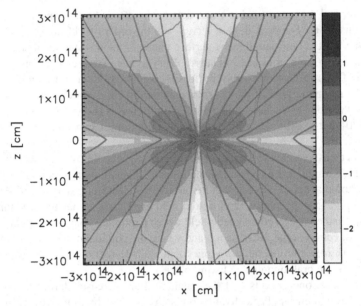

Fig. 9 (Color online) Magnetic field structure in a collapsing cloud core. The colour scale shows the intensity of the toroidal magnetic field component and the toroidal component is indicated by the *green lines*. The *blue line* shows the Alfvén surface of the outflow. The butterfly distribution of the toroidal field is typical for a magnetic field build up during the collapse of a rotating cloud core (from [8])

10 Summary

In summary one can conclude that outflows and jets can be driven very early during the collapse phase of self-gravitating objects. Self-consistent models show that during the collapse a large amount of magnetic flux is compressed into the protostellar region and a strong toroidal field component is generated in this process. Eventually the field configuration exerts a Lorentz force on the gas which is strong enough to lift material off the protostellar disc. Typically the earliest low velocity outflows are driven by magnetic pressure forces reminiscent of magnetic tower configurations in which the interior of the outflow is inflated by magnetic fields. Later into the collapse a jet is driven by magneto-centrifugal forces from the inner region of the protostellar object. The jet and outflow velocities are naturally linked to the depth of the gravitational well from which they are launched.

Whether these early outflows and jets are persistent or just transient features is one of the open issues in such a configuration. This question has to be perused in future research. But even then, this early type outflow play a central role in transporting disc angular momentum, and in determining the disc structure and basic properties of the forming protostars. Another open issue is how well the magnetically responding ions and the neutral gas are coupled. In the weakly coupled regime the magnetic field strength and geometry might be substantially different compared to the ideal MHD case discussed in this article. Two fluid calculations, where ions and neutrals are treated separately, or a realistic treatment of ambipolar diffusion is necessary to address these issues in the non-ideal MHD regime. In particular, it will be interesting to find out how much magnetic flux can be compressed into the protostar if effects from ambipolar diffusion are included in collapse calculations.

Acknowledgements I thank the JETSET Network which gave me the opportunity to present this lecture. I also thank the organisers of the 5th Jetset workshop for their hospitality. I am funded by the Deutsche Forschungsgemeinschaft within the Emmy-Noether grant BA 3706/1.

References

1. Allen, A., Li, Z., Shu, F.H., 2003, Collapse of Magnetized Singular Isothermal Toroids. II. Rotation and Magnetic Braking. ApJ 599, 363–379, DOI 10.1086/379243
2. Alves, J.F., Lada, C.J., Lada, E.A., 2001, Internal structure of a cold dark molecular cloud inferred from the extinction of background starlight. Nature 409, 159–161
3. Andre, P., Ward-Thompson, D., Barsony, M., 2000, From Prestellar Cores to Protostars: The Initial Conditions of Star Formation. Protostars and Planets IV pp 59–+
4. Arce, H.G., Shepherd, D., Gueth, F., Lee, C.F., Bachiller, R., Rosen, A., Beuther, H., 2007, Molecular Outflows in Low- and High-Mass Star-forming Regions. In: B. Reipurth, D. Jewitt, K. Keil (eds) Protostars and Planets V, pp 245–260
5. Bacciotti, F., Ray, T.P., Mundt, R., Eislöffel, J., Solf, J., 2002, Hubble Space Telescope/STIS Spectroscopy of the Optical Outflow from DG Tauri: Indications for Rotation in the Initial Jet Channel. ApJ 576, 222–231
6. Bachiller, R., 1996, Bipolar Molecular Outflows from Young Stars and Protostars. ARA&A 34, 111–154
7. Bally, J., Reipurth, B., Davis, C.J., 2007, Observations of Jets and Outflows from Young Stars. In: B. Reipurth, D. Jewitt, K. Keil (eds) Protostars and Planets V, pp 215–230
8. Banerjee, R., Pudritz, R.E., 2006, Outflows and jets from collapsing magnetized cloud cores. ApJ 641, 949–+
9. Banerjee, R., Pudritz, R.E., 2007, Massive star formation via high accretion rates and early disk-driven outflows. ApJ 660, 479, astro-ph/0612674
10. Banerjee, R., Pudritz, R.E., Holmes, L., 2004, The formation and evolution of protostellar discs; three-dimensional adaptive mesh refinement hydrosimulations of collapsing, rotating Bonnor-Ebert spheres. MNRAS 355, 248–272
11. Bate, M.R., Burkert, A., 1997, Resolution requirements for smoothed particle hydrodynamics calculations with self-gravity. MNRAS 288, 1060–1072
12. Beck, R., 2001, Galactic and Extragalactic Magnetic Fields. Space Sci Rev 99, 243–260, arXiv:astro-ph/0012402
13. Berger, M.J., Colella, P., 1989, Local adaptive mesh refinement for shock hydrodynamics. J Comput Phys 82, 64–84
14. Berger, M.J., Oliger, J., 1984, Adaptive mesh refinement for hyperbolic partial differential equations. J Comput Phys 53, 484–+
15. Beuther, H., Shepherd, D., 2005, Precursors of UCHII Regions and the Evolution of Massive Outflows. In: M.S.N. Kumar, M. Tafalla, P. Caselli (eds) Cores to Clusters: Star Formation with Next Generation Telescopes, pp 105–119
16. Blandford, R.D., Payne, D.G., 1982, Hydromagnetic flows from accretion discs and the production of radio jets. MNRAS 199, 883–903
17. Bonnor, W.B., 1956, Boyle's Law and gravitational instability. MNRAS 116, 351–+

18. Børve, S., Omang, M., Trulsen, J., 2006, Multidimensional MHD Shock Tests of Regularized Smoothed Particle Hydrodynamics. ApJ 652, 1306–1317, DOI 10.1086/508454

19. Boss, A.P., 2002, Collapse and Fragmentation of Molecular Cloud Cores. VII. Magnetic Fields and Multiple Protostar Formation. ApJ 568, 743–753, DOI 10.1086/339040

20. Braithwaite, J., Spruit, H.C., 2004, A fossil origin for the magnetic field in A stars and white dwarfs. Nature 431, 819–821, DOI 10.1038/nature02934

21. Brandenburg, A., Nordlund, A., Stein, R.F., Torkelsson, U., 1995, Dynamogenerated Turbulence and Large-Scale Magnetic Fields in a Keplerian Shear Flow. ApJ 446, 741–+, DOI 10.1086/175831

22. Cabrit, S., Raga, A., Gueth, F., 1997, Models of Bipolar Molecular Outflows. In: B. Reipurth, C. Bertout (eds) Herbig-Haro Flows and the Birth of Stars, IAU Symposium, vol 182, pp 163–180

23. Collins, D.C., Norman, M.L., 2004, Devolopment of an AMR MHD module for the code Enzo. In: Bulletin of the American Astronomical Society, Bullet Am Astron Soc 36, 1605–+

24. Contopoulos, J., 1995, A simple type of magnetically driven jets: An astrophysical plasma gun. ApJ 450, 616–+, DOI 10.1086/176170

25. Contopoulos, J., 1996, General Axisymmetric Magnetohydrodynamic Flows: Theory and solutions. ApJ 460, 185–+, DOI 10.1086/176960

26. Crutcher, R.M., Troland, T.H., Lazareff, B., Paubert, G., Kazès, I., 1999, Detection of the CN Zeeman Effect in Molecular Clouds. ApJ 514, L121–L124

27. Desch, S.J., Mouschovias, T.C., 2001, The magnetic decoupling stage of star formation. ApJ 550, 314–333

28. Donati, J.F., Paletou, F., Bouvier, J., Ferreira, J., 2005, Direct detection of a magnetic field in the innermost regions of an accretion disk. Nature 438, 466–469

29. Duffin, D.F., Pudritz, R.E., 2008, Simulating hydromagnetic processes in star formation: introducing ambipolar diffusion into an adaptive mesh refinement code. MNRAS in press arXiv:0810.0299

30. Ebert, R., 1955, "Uber die Verdichtung von H I-Gebieten. Mit 5 Textabbildungen. Zap 37, 217–+

31. Fendt, C., Camenzind, M., 1996, Magnetohydrodynamic Structure of Protostellar Jets. Astrophys Lett Commun 34, 289–+

32. Ferreira, J., 1997, Magnetically-driven jets from Keplerian accretion discs. A&A 319, 340–359

33. Foster, P.N., Chevalier, R.A., 1993, Gravitational collapse of an isothermal sphere. ApJ 416, 303–+

34. Fromang, S., Hennebelle, P., Teyssier, R., 2005, RAMSES-MHD: an AMR Godunov code for astrophysical applications. In: F. Casoli, T. Contini, J.M. Hameury, L. Pagani (eds) SF2A-2005: Semaine de l'Astrophysique Francaise, pp 743–+

35. Fryxell, B., Olson, K., Ricker, P., Timmes, F.X., Zingale, M., Lamb, D.Q., MacNeice, P., Rosner, R., Truran, J.W., Tufo, H., 2000, FLASH: An adaptive mesh hydrodynamics code for modeling astrophysical thermonuclear flashes. ApJS 131, 273–334

36. Hennebelle, P., Fromang, S., 2008, Magnetic processes in a collapsing dense core. I. Accretion and ejection. A&A 477, 9–24, DOI 10.1051/0004-6361:20078309, arXiv:0709.2886

37. Hosking, J.G., Whitworth, A.P., 2004, Fragmentation of magnetized cloud cores. MNRAS 347, 1001–1010, DOI 10.1111/j.1365-2966.2004.07274.x

38. Hosking, J.G., Whitworth, A.P., 2004, Modelling ambipolar diffusion with two-fluid smoothed particle hydrodynamics. MNRAS 347, 994–1000, DOI 10.1111/j.1365-2966.2004.07273.x

39. Johns-Krull, C.M., Valenti, J.A., Koresko, C., 1999, Measuring the magnetic field on the classical T Tauri Star BP Tauri. ApJ 516, 900–915

40. Kato, Y., Mineshige, S., Shibata, K., 2004, Magnetohydrodynamic accretion flows: Formation of magnetic tower jet and subsequent quasi-steady state. ApJ 605, 307–320

41. Königl, A., Pudritz, R.E., 2000, Disk winds and the accretion-outflow connection. Protostars and Planets IV pp 759–+

42. Lada, C.J., Alves, J.F., Lombardi, M., 2007, Near-Infrared Extinction and Molecular Cloud Structure. In: B. Reipurth, D. Jewitt, K. Keil (eds) Protostars and Planets V, pp 3–15
43. Larson, R.B., 1969, Numerical calculations of the dynamics of collapsing protostar. MNRAS 145, 271–+
44. Levy, E.H., Sonett, C.P., 1978, Meteorite magnetism and early solar system magnetic fields. In: IAU Colloq. 52: Protostars and Planets, pp 516–+
45. Li, P.S., Norman, M.L., Mac Low, M., Heitsch, F., 2004, The formation of self-gravitating cores in turbulent magnetized clouds. ApJ 605, 800–818
46. Lubow, S.H., Papaloizou, J.C.B., Pringle, J.E., 1994, Magnetic field dragging in accretion discs. MNRAS 267, 235–240
47. Lynden-Bell, D., 2003, On why discs generate magnetic towers and collimate jets. MNRAS 341, 1360–1372
48. Machida, M.N., Tomisaka, K., Matsumoto, T., 2004, First MHD simulation of collapse and fragmentation of magnetized molecular cloud cores. MNRAS 348, L1–L5
49. Machida, M.N., Matsumoto, T., Hanawa, T., Tomisaka, K., 2005, Collapse and fragmentation of rotating magnetized clouds - II. Binary formation and fragmentation of first cores. MNRAS 362, 382–402, DOI 10.1111/j.1365-2966.2005.09327.x, arXiv:astro-ph/0506440
50. Machida, M.N., Matsumoto, T., Tomisaka, K., Hanawa, T., 2005, Collapse and fragmentation of rotating magnetized clouds – I. Magnetic flux-spin relation. MNRAS 362, 369–381, DOI 10.1111/j.1365-2966.2005.09297.x, arXiv:astroph/0506439
51. Machida, M.N., Inutsuka, Si., Matsumoto, T., 2006, Outflows driven by giant protoplanets. ApJ 649, L129–L132, DOI 10.1086/508256, arXiv:astroph/0604594
52. Machida, M.N., Matsumoto, T., Hanawa, T., Tomisaka, K., 2006, Evolution of Rotating Molecular Cloud Core with Oblique Magnetic Field. ApJ 645, 1227–1245, DOI 10.1086/504423, arXiv:astro-ph/0602034
53. Machida, M.N., Inutsuka, Si., Matsumoto, T., 2007, Magnetic Fields and Rotations of Protostars. ApJ 670, 1198–1213, DOI 10.1086/521779, arXiv:astroph/0702183
54. MachidaMN, Inutsuka, Si., Matsumoto, T., 2008, High- and Low-VelocityMagnetized Outflows in the Star Formation Process in a Gravitationally Collapsing Cloud. ApJ 676, 1088–1108, DOI 10.1086/528364
55. Matsumoto, T., Tomisaka, K., 2004, Directions of outflows, disks, magnetic fields, and rotation of ysos in collapsing molecular cloud cores. ApJ 616, 266–282, astro-ph/0408086
56. Meglicki, Z., 1994, Verification and accuracy of smoothed particle magnetohydrodynamics. Comput Phys Commun 81, 91–104, DOI 10.1016/0010-4655(94)90113-9
57. Mignone, A., Bodo, G., Massaglia, S., Matsakos, T., Tesileanu, O., Zanni, C., Ferrari, A., 2007, PLUTO: A Numerical Code for Computational Astrophysics. ApJS 170, 228–242, DOI 10.1086/513316, arXiv:astro-ph/0701854
58. Mouschovias, T.C., Paleologou, E.V., 1979, The angular momentum problem and magnetic braking - an exact time-dependent solution. ApJ 230, 204–222
59. Mouschovias, T.C., Paleologou, E.V., 1980, Magnetic braking of an aligned rotator during star formation – an exact, time-dependent solution. ApJ 237, 877–899
60. O'Shea, B.W., Bryan, G., Bordner, J., Norman, M.L., Abel, T., Harkness, R., Kritsuk, A., 2004, Introducing Enzo, an AMR Cosmology Application. ArXiv Astrophysics e-prints astro-ph/0403044
61. Ouyed, R., Pudritz, R.E., 1997, Numerical Simulations of Astrophysical Jets from Keplerian Disks. I. Stationary Models. ApJ 482, 712–+, DOI 10.1086/304170
62. Price, D.J., Monaghan, J.J., 2004, Smoothed Particle Magnetohydrodynamics – I. Algorithm and tests in one dimension. MNRAS 348, 123–138, DOI 10.1111/j.1365-2966.2004.07345.x, arXiv:astro-ph/0310789
63. Price, D.J., Monaghan, J.J., 2004, Smoothed ParticleMagnetohydrodynamics – II. Variational principles and variable smoothing-length terms.MNRAS 348, 139–152, DOI 10.1111/j.1365-2966.2004.07346.x, arXiv:astro-ph/0310790

64. Price, D.J., Monaghan, J.J., 2005, Smoothed Particle Magnetohydrodynamics – III. Multi-dimensional tests and the $\nabla.B = 0$ constraint. MNRAS 364, 384–406, DOI 10.1111/j.1365-2966.2005.09576.x, arXiv:astro-ph/0509083

65. Pudritz, R.E., 1981, Dynamo action in turbulent accretion discs around black holes – part two – the mean magnetic field. MNRAS 195, 897

66. Pudritz, R.E., 1981, Dynamo action in turbulent accretion discs around black holes. I – The fluctuations. II – The mean magnetic field. MNRAS 195, 881

67. Pudritz, R.E., 2003, Accretion-Ejection Models of Astrophysical Jets. NATO ASI, Les Houches, Session LXXVIII, *Accretion Discs, Jets and High Energy Phenomena in Astrophysics* pp p. 187–230

68. Pudritz, R.E., Norman, C.A., 1983, Centrifugally driven winds from contracting molecular disks. ApJ 274, 677–697

69. Pudritz, R.E., Ouyed, R., Fendt, C., Brandenburg, A., 2007, Disk Winds, Jets, and Outflows: Theoretical and Computational Foundations. In: B. Reipurth, D. Jewitt, K. Keil (eds) Protostars and Planets V, pp 277–294

70. Pudritz, R.E., Banerjee, R., Ouyed, R., 2008, The role of jets in the formation of planets, stars, and galaxies. In: Charbrier, G. (ed) Structure formation in the Universe

71. Reipurth, B., Bally, J., 2001, Herbig-Haro Flows: Probes of Early Stellar Evolution. ARA&A 39, 403–455, DOI 10.1146/annurev.astro.39.1.403

72. Ruffert, M., 1992, Collisions between a white dwarf and a main-sequence star. II – Simulations using multiple-nested refined grids. A&A 265, 82–105

73. Shepherd, D.S., Churchwell, E., 1996, Bipolar molecular outflows in massive star formation regions. ApJ 472, 225–+, DOI 10.1086/178057

74. Shepherd, D.S., Churchwell, E., 1996, High-velocity molecular gas from high-mass star formation regions. ApJ 457, 267–+, DOI 10.1086/176727

75. Shibata, K., Uchida, Y., 1985, A magnetodynamic mechanism for the formation of astrophysical jets. I – Dynamical effects of the relaxation of nonlinear magnetic twists. PASJ37, 31–46

76. Stepinski, T.F., Levy, E.H., 1988, Generation of dynamo magnetic fields in protoplanetary and other astrophysical accretion disks. ApJ 331, 416–434, DOI 10.1086/166569

77. Stone, J.M., Gardiner, T.A., Teuben, P., Hawley, J.F., Simon, J.B., 2008, Athena: A new code for astrophysical MHD. ArXiv e-prints 804, 0804.0402

78. Tomisaka, K., 1998, Collapse-driven outflow in star-forming molecular cores. ApJ 502, L163+

79. Tomisaka, K., 2002, Collapse of rotating magnetized molecular cloud cores and mass outflows. ApJ 575, 306–326

80. Truelove, J.K., Klein, R.I., McKee, C.F., Holliman, J.H., Howell, L.H., Greenough, J.A., 1997, The jeans condition: A new constraint on spatial resolution in simulations of isothermal self-gravitational hydrodynamics. ApJ 489, L179+

81. Uchida, Y., Shibata, K., 1985, Magnetodynamical acceleration of CO and optical bipolar flows from the region of star formation. PASJ37, 515–535

82. von Rekowski, B., Brandenburg, A., Dobler, W., Dobler, W., Shukurov, A., 2003, Structured outflow from a dynamo active accretion disc. A&A 398, 825–844, DOI 10.1051/0004-6361:20021699

83. Wang, P., Abel, T., 2009, Magnetohydrodynamic simulations of disk galaxy formation: the magnetization of the cold and warm medium. ApJ 696, 96–109

84. Xu, H., Collins, D.C., NormanML, Li, S., Li, H., 2008, A cosmological AMRMHD module for Enzo. ArXiv e-prints 804, 0804.1334

85. Yorke, H.W., Bodenheimer, P., Laughlin, G., 1993, The formation of protostellar disks. I – 1 M(solar). ApJ 411, 274–284

86. Ziegler, E., Dolag, K., Bartelmann, M., 2006, Divergence cleaning techniques in smoothed particle magnetohydrodynamics simulations. Astronomische Nachrichten 327, 607–+, DOI 10.1002/asna.200610602

87. Ziegler, U., 2005, Self-gravitational adaptive mesh magnetohydrodynamics with the NIRVANA code. A&A 435, 385–395, DOI 10.1051/0004-6361:20042451

Index

Gracia, J. et al.: *Index*. Lect. Notes Phys. **791**, 223–227 (2009)
DOI 10.1007/978-3-642-03370-4   © Springer-Verlag Berlin Heidelberg 2009